HIDDEN HUNGER

HIDDEN HUNGER

Gender and the Politics of
Smarter Foods

Aya Hirata Kimura

CORNELL UNIVERSITY PRESS   **ITHACA AND LONDON**

Cornell University gratefully acknowledges receipt of a grant from the Women's Studies Department of the University of Hawai'i, which assisted in the publication of this book.

First published 2013 by Cornell University Press
First printing, Cornell Paperbacks, 2013

Printed in the United States of America

Library of Congress Cataloging-in-Publication Data

Kimura, Aya Hirata, 1974–
    Hidden hunger : gender and the politics of smarter foods / Aya Hirata Kimura.
        p. cm.
    Includes bibliographical references and index.
    ISBN 978-0-8014-5164-5 (cloth : alk. paper) —
    ISBN 978-0-8014-7859-8 (pbk. : alk. paper)
    1. Nutrition policy—Indonesia. 2. Women—Nutrition—Indonesia.
3. Malnutrition—Indonesia—Prevention. 4. Enriched foods—Indonesia. 5. Trace elements in nutrition—Indonesia. 6. Food habits—Political aspects—Indonesia.
I. Title.
    TX360.I5K56   2013
    362.1963'9009598—dc23   2012029708

Frederick H. Buttel (1948–2005)

Koyoshi Nakano (1917–1999)

Contents

Tables and Figures

Tables

Figures

Acknowledgments

This book has come about through the encouragement of many people. At the University of Wisconsin–Madison, Frederick H. Buttel gave enthusiastic support for the project. Unfortunately, he died while I was doing my fieldwork in Indonesia. This book is dedicated Fred, who was not only a brilliant scholar but also a true teacher. Jane Collins has been a great mentor who always provides thoughtful suggestions and advice. The book would not have been possible without rich conversations and guidance from Jack Kloppenburg Jr., Samer Alatout, Daniel Lee Kleinman, and Clark Miller. Other students and faculty members in the departments of Rural (now Community and Environmental) Sociology and Sociology, and the Center for Southeast Asian Studies, provided a supportive atmosphere, friendship, and companionship.

I also appreciate the willing assistance of the many Indonesian researchers, policymakers, and NGO workers with whom I spoke. Pak Soekirman kindly shared his knowledge of nutritional policy in the country as well as his vast social network, which was critical for my research in Indonesia. Professors Aman Wirakartakusumah and Adil Ahza at Bogor Agricultural University in Bogor gave institutional support. Nelden and Yosef Djakababa's kindness, hospitality, and friendship were key to my survival during the fieldwork. I am also grateful to people whom I interviewed in the United States about their work in international nutrition and development, including Alfred Summer and staff members at the World Bank, USAID, the International Life Sciences Institute, and the International Food Policy Research Institute.

I also benefited from the intellectual and personal support from my colleagues at the Women's Studies Department at the University of Hawai'i–Manoa. In particular, Susan Hippensteele was wonderfully welcoming to me when I first came to Hawaii. Kathy Ferguson was generous with her time and read early drafts of the book. Her thoughtful comments helped me to articulate the gendered dimensions of food politics. Meda Chesney-Lind provided indispensable support for the project as chair of the Women's Studies Department.

I also appreciate the encouragement from Phil McMichael and Michael Dove, and feedback from Christine Yano, Jane Freeman Moulin, and Pensri Ho, who read early drafts of the book. Carol Colfer at the Center for International Forestry Research and an anonymous reviewer for Cornell University Press offered insightful comments. Roger Haydon, executive editor at Cornell University Press, skillfully guided me through the overall project. Special thanks are also due to Ange Romeo-Hall, senior manuscript editor, and Katy Meigs, copy editor, at Cornell University Press for their helpful comments and thorough editing. This research was financially supported by grants from the National Science Foundation, the Rural Sociological Society, and the Center for Southeast Asian Studies at the University of Wisconsin–Madison.

I wish to thank my family as well. My mother, Ryoko (Nakano) Hirata, and my father, Masahiro Hirata, always let me pursue my dreams with strong faith and love. I owe my deepest gratitude to my partner, Ehito Kimura, who has endured all stages of this project with a big heart and soul. Our children, Isato and Emma, remind me even in the most mundane way that all children deserve a world without hunger. Their great grandmother, Koyoshi Nakano, to whom I also dedicate the book, raised generations of confident and empathetic women and inspired me by her resilience and strength.

Finally, I am deeply grateful to the many Indonesian women who let me into their homes and shared their mothering stories. Now that I am a mother of two, I consider what they had to tell me with deeper appreciation. It is my hope that these mothers find the book reflects their experiences.

Abbreviations

ACC/SCN UN	Administrative Committee on Coordination, Sub-Committee on Nutrition
ADB	Asian Development Bank
APTINDO	Indonesian Association of Wheat Flour Producers
BAFF	Business Alliance for Food Fortification
BAPPENAS	National Development Planning Board, Indonesia
BIMAS	Mass Guidance program, Indonesia
BKKBN	National Family Planning Coordinating Board
BULOG	Food Logistics Agency, Indonesia
CF	complementary food
CGIAR	Consultative Group on International Agricultural Research
CYMMIT	International Maize and Wheat Improvement Center
DALYs	disability adjusted life years
FAO	UN Food and Agriculture Organization
GAIN	Global Alliance for Improved Nutrition
GMO	genetically modified organism
HKI	Helen Keller International
HNP	Health, Nutrition and Population
ICN	International Conference on Nutrition
IDA	iron deficiency anemia
IDD	iodine deficiency disorder
IFPRI	International Food Policy Research Institute
ILSI	International Life Sciences Institute
IRRI	International Rice Research Institute
ISAAA	International Service for the Acquisition of Agri-biotech Applications
IVACG	International Vitamin A Consultative Group
KFI	Indonesian Fortification Coalition (Koalisi Fortifikasi Indonesia)
MDG	Millennium Development Goal
MI	Micronutrient Initiative
OMNI	Opportunities for Micronutrient Initiatives
Persagi	Indonesian Nutritionist Association (Persatuan Ahli Gizi Indonesia)

PAMM	Program Against Micronutrient Malnutrition
PATH	Program for Appropriate Technology for Health
PAG	Protein Advisory Group
RDA	recommended daily allowance
Repelita	Five-Year Development Plan (Rencana Pembangunan Lima Tahun)
SAP	Structural Adjustment Program
SF	supplementary food
SKRT	National Household Health Survey, Indonesia
SNI	Indonesian National Standard
SUSENAS	National Social Economic Survey, Indonesia
UNICEF	United Nations Children's Fund
UPGK	Family Nutrition Improvement Program (Usaha Perbaikan Gizi Keluarga)
USAID	US Agency for International Development
VAD	vitamin A deficiency
WFP	UN World Food Programme
WHO	UN World Health Organization

UNCOVERING HIDDEN HUNGER

Obviously, what hungry people need first and foremost is more food.
But they also need *better* food.

—*Economist,* July 31, 2004

One of the great Western misconceptions is that severe malnutrition
is simply about not getting enough to eat. Often it's about not getting the right micronutrients—iron, zinc, vitamin A, iodine—and one
of the most cost-effective ways outsiders can combat poverty is to
fight this "hidden hunger."

—Nicholas Kristof, *New York Times,* May 24, 2009

Shiny red and blue packages of cookies and instant noodles replete with appetizing
photos and fancy logos arrived at a cluster of small shacks that constitute a tiny
portion of the vast Jakarta slums. Mothers took the noodles for themselves and
the cookies for their children. Although they resemble common junk food,
these products are actually healthy foods according to the UN World Food Programme. They are fortified with iron, zinc, calcium, magnesium, phosphorus,
potassium, vitamins A, D, E, K, B_1, B_2, B_6, and B_{12}, and folic acid. The WFP's
enthusiasm for fortified foods is shared by the government of Indonesia, which
decided on mandatory wheat flour fortification in 1998 and began distributing
fortified baby food to low-income families in 2001. The baby food was fortified
with iron, zinc, calcium, magnesium, potassium, vitamins A, D, K, B_1, B_2, and
B_{12}, and folic acid.

In the 1990s, the lack of proper micronutrients—or "micronutrient deficiency"—became a hot topic in the international food policy community to
describe the "food problem" in the developing world. A previously hidden, yet
deadly, aspect of the condition of Third World people, micronutrient deficiency,
or "hidden hunger," became the focus of many development projects. The term
"micronutrients" refers to vitamins and minerals that are vital for the proper
functioning of the body; examples of micronutrient deficiencies include vitamin A deficiency, iron deficiency anemia, and iodine deficiency disorder. These
disorders are often not apparent to the people with a deficiency, hence it is
called hidden hunger. In the 1990s, many international conferences, from the

World Summit for Children to the World Food Summit, urged governments to recognize the importance of micronutrient deficiencies, and many international and philanthropic donors started to commit resources to combating hidden hunger. The degree to which concern with micronutrient deficiencies became established within the development discourse could be seen, for instance, in the 2008 Copenhagen Consensus conference of leading international economists and specialists, who chose micronutrient remediation as one of the most cost-effective development interventions. In another example from around the same time, a well-known pioneer in microfinance, the Grameen Bank, formed a joint venture with the French multinational corporation Danone/Dannon to produce fortified yogurt in Bangladesh in 2006.[1]

There are various policies to address hidden hunger, but fortification and bio-fortification, not supplements or nutrition education, became the most celebrated instruments for addressing it in the final decade of the twentieth century. "Forti-fication" refers to the process of adding micronutrients to food products during the manufacturing process, as when vitamins are added to baby food, wheat flour, sugar, cooking oil, or butter. Indonesian fortified cookies and instant noodles are examples. "Biofortification" alters crops biologically so that the plants themselves contain more micronutrients; the prime example of this is genetically engineered Golden Rice, which has enhanced beta-carotene, a vitamin A precursor. Both for-tification and biofortification are responses to concerns about the micronutrient intake of the poor.

In this book I explore the politics of the recent turn to micronutrients by examining international projects and agreements on hidden hunger as well as by using case studies from Indonesia. The Indonesian cases illustrate how micro-nutrient deficiencies gained prominence in expert discourse in the 1990s and interest in fortification and biofortification increased, despite hunger and limited access to sufficient quantity of food still being rampant in some communities. I will show, for instance, how mandatory fortification was started with wheat flour, and how Golden Rice was promoted by biofortification proponents to Indonesia's policymakers in the 1990s. It is tempting to portray this interest in "quality" and hidden hunger as driven solely by the latest advances in nutritional science. The increased attention to hidden hunger might be viewed as the result of scientific progress uncovering previously hidden human needs, now revealed as micronutrient deficiencies. This might be seen as the logical extension of how the hunger problem is viewed since the impressive increase in global food pro-duction through the application of modern technologies.

This view creates several puzzles, however. First, why was it that the 1990s saw the micronutrient turn, even though scientists had known about micronutrients and their health implications for over half a century? Since the early twentieth century, the functions of micronutrients have been recognized, and fortification has been implemented in developed countries.[2] Furthermore, the preferred solution to the problem—fortification and biofortification—substantially diverges from what many activists and scholars have advocated as the best way to achieve a sustainable, secure, and stable food system. Biofortification uses controversial genetically engineered crops. Fortification, by its very nature, depends on processed foods, which some have criticized for distorting traditional dietary patterns and increasing the potential for chronic diseases. The emphasis on fortification also leads to a lucrative business opportunity for multinational companies. In this book I explore how and why fortification and biofortification became the preferred "solutions" to the Third World food problem. Tracing trends in the international development discourse and through detailed cases of three categories of standard micronutrient-oriented programs (mandatory fortification, voluntary fortification, and biofortification), I show how activities related to fortification and biofortification of micronutrients increased. I believe that this micronutrient turn was driven by "nutritionism" and that it ought to be understood as a manifestation of a scientized view of food insecurity in developing countries.

Nutritionism

While attention to the nutritional quality of food might be considered a welcome change from an earlier focus on food quantity, I suggest that when it is driven by "nutritionism," it has serious political and gendered implications. Nutritionism refers to an increasingly prevalent view that food is primarily a vehicle for delivering nutrients. Gyorgy Scrinis (2008, 39) defines it as "nutritionally reductive approach to food" that "has come to dominate, to undermine, and to replace other ways of engaging with food and of contextualizing the relationship between food and the body." The goodness of food depends on the type and amount of nutrients. Health improvement becomes the foremost purpose of food and of the act of eating.[3]

Nutritionism is so pervasive that it is often hard to notice how peculiar it is. But it is highly reductionist. Food and eating have layered meanings and values that go well beyond nutritional properties and contributions to physical well-being.[4] A list of nutrients, however comprehensive, cannot capture the richness of the cultural, social, and historical meanings of food that are intimately tied to family, community, and ethnicity and, as well, to social status and power. Additionally, pleasure, not only wellness, can be the objective of eating. People eat for various reasons, and the discourse of health and nutrition captures only one dimension of the act of eating.

Nutritionism is often understood as a kind of marketing gimmick in the sophisticated consumer market of the global North. Michael Pollan, who has written several popular books on US food politics, has explained how the concept of nutritionism enabled food companies to market processed foods as "healthy" food, resulting in an increase in obesity in the United States (2008). With so many "functional foods" and "nutraceuticals" flooding the supermarket shelves, it is not difficult to see why nutritionism's theorization has so far been focused on industrial nations. But nutritionism has become influential globally. "Smart foods," or food fortified with added vitamins and minerals for enhancing functional benefits, are no longer the monopoly of health-conscious shoppers in developed countries. They are now a part of antihunger and antimalnutrition strategies in developing countries.

Furthermore—and here I follow anthropologists of international development who locate projects to improve the welfare of people in the global South in the field of governmentality—I situate nutritionism as a technique of power.[5] There is no doubt that nutritionism creates profitable marketing opportunities for food companies. But nutritionism is also tied to new modes of governance and consciousness and subjectivity of individuals that are particularly compatible with the neoliberal age. Placing nutritionism within the complex relations of power-knowledge (Foucault 1980), I argue that nutritionism is part and parcel of the long history of problematizing people's food and bodies in the developing world, particularly through the deployment of modern scientific and technical expertise. In her analysis of projects driven by the "will to improve" in Indonesia, Tania Li (2007) discusses how such projects require "the practice of 'rendering technical'" that makes contentious issues a delimited technical matter. The result is depoliticization as well as a boundary between those "with the capacity to diagnose deficiencies in others and those who are subject to expert direction" (7–21). Nutritionism follows this long-standing practice of improvement schemes by "benevolent and stubborn trustees" who "claim to know how others should live, to know what is best for them, to know what they need" (4).

I chart four important dimensions of nutritionism in the context of Third World food politics. First is the rise of what I term *charismatic nutrient* and corresponding *nutritional fixes,* technical attempts to solve the Third World food problem that target only its nutritional aspect.[6] Because nutritionism marks the problem of Third World food as chemical and individual, it follows that the Third World food problem is essentially the problem of "inferior" food. The poorness of particular diets is calculated based on the discrepancy between an individual's intake of nutrients and scientifically set standards. The way to correct a bad diet is to provide the essential missing nutrients in the most efficient form for delivery, be it a pill, fortified cookies, or a biofortified crop. As we will see, different charismatic nutrients have been celebrated as the key to combating the Third World food problem at different historical periods, and various "solutions"—nutritional fixes in a different guise—have been proposed based on this reductionist understanding of the food problem.

A second dimension of nutritionism is that it effectively depoliticizes the food problem by recasting it as a technical matter. Nutritionism tends to individualize the Third World food problem by adopting chemically analyzable nutrient makeup and biochemical parameters as standards for measuring the health of food and bodies. By creating a discursive field of identifiable missing nutrients, nutritionism refashions the food problem. Food problems become a matter of individual self-discipline, of "awareness" and "behavior," with corresponding market-based solutions. One critical consequence of such framing is that it fits the food problem inside increasingly precise nutritional parameters, removing other ways of discussing it. Nutritional composition of food and bad eating habits of individuals come to be considered the problem, rather than living conditions, low wages, lack of land and other productive resources, or rising food prices. By profoundly limiting the frame of analysis and the usable vocabulary, nutritionism critically shapes the construction of the food problem and limits the range of possible conversations.

Third, nutritionism in food security policies is shaped by larger development discourses, and the micronutrient turn in the 1990s was inseparable from overall neoliberalization. Unlike other mechanisms to address micronutrient deficiencies, such as nutrition education and supplement distribution, typically done by governments and/or international organizations, fortification and biofortification are more market driven and efficient alternatives. Although governmental agencies could implement programs, often the expertise necessary (such as intellectual property rights, manufacturing and marketing know-how, and so on) is held by private industry, and vitamins are added to existing food products made by private companies, so that fortification and biofortification are celebrated as instances of public-private partnership.[7] On another level, the interest in micronutrients coincided with a decrease in public funding for international agricultural research. In the 1980s, the productivist paradigm that had dominated international development started to fall out of favor. Green Revolution programs were funded and supported by governments and dependent on subsidized seeds, fertilizer, water, and other agricultural infrastructure.[8] But after the 1980s, governments increasingly disengaged from international agricultural projects, and international agricultural research centers also suffered from major funding cuts. The proportion of agricultural research done by the private sector increased, with an emphasis on inventions that were amenable to patent protection (Alston, Dehmer, and Pardey 2006). In this way, the micronutrient turn of the 1990s was profoundly shaped by neoliberalization, which, on the one hand, propelled the retreat of government from agricultural policy and, on the other, saw fortification and biofortification as market-based programs.[9]

Fourth, nutritionism critically shifts who can speak authoritatively about the food problem and who is listened to. In emphasizing the technical nature of the problem and solution, nutritionism privileges experts over lay people. This is particularly evident with "hidden hunger," where "patients" may not be aware that they have a problem. By legitimizing the domination of experts, nutritionism circumvents democratic processes in contemporary food politics. Nutritionism closes rather than expands avenues for citizen dialogue and participation in the making of better food systems. In the world of nutritionism, people credentialed as experts—not the poor women who are mainly responsible for feeding families and who also suffer from micronutrient deficiencies—are the ones who "know" the problem and hence can prescribe solutions for the malnourished. Conversations about food and food security in the Third World are filled with the claims and counterclaims of experts, but the silence of women who make food every day is a serious issue. It is precisely the voices of these women, who can describe the lived realities of malnutrition and hunger, that we need to make audible if we are to understand food's political and social, not simply its nutritional and medical, meanings.

Nutritionism systematically organizes knowledge about food and bodies, privileging an expert view while silencing other views. Nutritional science not only provides new knowledge and insight into the relationships between health and nutrients, it also fashions vocabularies for talking about food. By privileging academic credentials and public health contributions, nutritionism sets the parameters of acceptable debate. As we will see in Indonesia, nutritionism profoundly shapes how experts actually act on food and bodies in the Third World.

Feminist Food Studies

If the study of food has only recently begun to earn academic legitimacy, feminist food studies are of even more recent origin. As Avakian and Haber (2005) note, despite the long historical and cultural associations between women and food, only recently has a feminist perspective been brought to the study of food. That it is mostly women who produce and prepare food and feed people has been ignored or taken for granted.[10] Food is profoundly gendered. Throughout the world, women are primarily responsible for the purchasing and cooking of food, and they have a central role in the allocation of food that impacts the nutritional status of family members. Women play an important role in food production,[11] and an increasing number of women are employed globally to produce food.[12] Transnational corporations profit by paying women in developing countries lower wages, which is justified by the idea of women's "natural" skill in handling fragile food products (Collins 1995; Raynolds 1998), and made possible in part by "family obligations" that force them to accept temporary and seasonal work (Raynolds 1998). With the push for export-driven agriculture, contract labor has expanded in developing countries, also taking advantage of the flexible skilled labor of family members, especially women.[13]

Employment in the global agrofood economy is an important part of the relationship between women and food, but it is not the whole story. As food is an important component of any vision of a nation's development, people's well-being, and the stability of international and national order, food and agriculture have been on the agenda of many governments and international organizations. Policy interventions into food and nutrition are prevalent and often provoke much political ardor; their impacts on culture, economy, and the lives of women in developing countries are undertheorized. Of course, we have a lot of writing about how food policies ought to be reformed and improved. We have good accounts of failed state food policies and resulting famines, humanitarian crises, and hunger.[14] But the intersection of gender and food policy still produces many unanswered questions. To what extent do women have power to shape food and nutrition policies? How does gender ideology intersect with the state's aspiration to control food and bodies?

These questions are critical, for food policies often have contradictory implications for women. Historians have noted that efforts to improve food situations have tended to attract many women as active players, giving them social recognition and opportunities to enter a previously closed public domain.[15] On the other hand, food policies often have interacted powerfully with conservative social ideologies. "Unattractive" and "ill-cooked" meals made by women have been criticized as the source of social ills ranging from labor upheavals (Levenstein 1993) to alcoholism (Shapiro 2009).[16] Hence *commendation* of women's role in improving food has often been coupled with *condemnation* of women for not fulfilling their familial, nationalistic, and humanistic duties.

Contemporary food policies also bring a peculiar visibility to women. In many writings on the Third World food problem, women surface as a solution that celebrates their role in food reform. But often women are considered the solution because their inadequacy is the problem to be rectified. From governments' and experts' perspectives, women's food knowledge, cooking ability, feeding practices, and breast-feeding patterns are the means to solve the food problem, precisely because they are the origin of that problem. In this sense, women's visibility is rooted in committing a sin and providing a solution to rectify it.

One of the key arguments of this book is how discourses of the Third World food problem identify women, particularly mothers, as the key site of state policing and surveillance. It is worth pondering the parallel between the population issue and food. Feminist scholars have pointed to intense state interest in demographic changes, population control, and reproductive issues and how these interests have brought women's bodies increasingly under surveillance and control by governments and experts.[17] Through a demographic lens, women's bodies are linked with national and global futures (Gupta 2001; Unnithan-Kumar 2004). Food is much like population in being invested in modernity and national development and also with transnational anxiety over geopolitical stability. States and international development organizations assert that food—like people—is an important ingredient in "modernizing" and "developing" the Third World.[18] With their longstanding association with food, cooking, and feeding, women are implicitly and explicitly targeted by the state and development organizations and scientific experts.

Simultaneously, women's peculiar visibility in food reform is situated in a capitalist food system. Posing as a partner in food reform, the food industry is rarely an outside observer of movements to improve food. Capitalizing on the anxieties of women has been a mainstay of its marketing strategies. Fears about alienated husbands, disappointed children, and embarrassed guests often figure prominently in advertisements that also offer commoditized solutions (Parkin 2006). Mothers are a supreme target of commercial advertisements for products from educational materials to baby food, transforming child rearing into what scholars have called "consuming motherhood" (Taylor, Layne, and Wozniak 2004). In short, both scientization and commodification shape contemporary food policies, staging women as both the solution and the culprit.

Furthermore, nutritionism accords women a visibility in another limited framework, that of *biological victimhood*. While broad discussion about women's nutritional status is a staple of contemporary food and nutrition policies, such talk brings women onto the horizon of policy debates primarily as abstract members of a biologically determined group. Rooted in mainstream nutritional science's embrace of quantifiable biological indicators of human nutritional status, nutritionism takes women as a homogeneous group with a shared biological identity and codes them with a biological propensity to nutritional diseases.

The chapters to follow provide empirical evidence for the gendered nature of food policies in developing countries. We will see how nutritionism creates a particular visibility for women—but not necessarily in a way that reduces their oppression and marginality. Discourses of nutritionism may highlight women's plight as the victims of micronutrient deficiencies, but only as biologically programmed ones. Women's food may be recognized as an important factor in shaping the nutritional status of the population, and experts and companies may celebrate women's role in providing optimum nutrition. Yet despite this celebration of women's role, optimum nutrients and profits, not the optimum situation for individual women, are the core concerns for experts and companies. Women are simultaneously victim, savior, and culprit. Such gendered liabilities are critical in constituting the contemporary relationship between food and women in developing countries.

Theoretical Contexts

In addition to feminist studies, this book is in conversation with what is often called agrofood studies and science and technology studies. Over the past several decades, agrofood studies have made significant efforts to understand the politics of food and agriculture around the world. Agrofood scholars have examined the political economy of food; the history of the industrialization of agriculture and its geopolitical structure; and the ecological, social, and cultural consequences of a changing agrofood system.[19] I share the concerns of many in the field for sustainable and socially just food systems.

One of the key contributions of agrofood studies has been to politicize the understanding of antihunger, antimalnutrition programs and to explore the political and social structuring of interventions into food systems in the developing world; these interventions are often concealed by humanitarian framing. In her brilliant analysis of historical shifts in the global food system, Harriet Friedmann analyzed food aid to the developing countries as a critical component of what she and Phil McMichael call "the second food regime" (Friedmann and McMichael 1989), which enabled the United States to dispose of surplus grains.[20] Pressed to deal with agricultural surpluses accumulated through government purchases that aimed to raise agricultural prices, the US government created Public Law 480 (the Food for Peace program) in 1954 and started dispensing surplus wheat to developing countries.[21] Food aid came to constitute a substantial portion of the total world trade in wheat by the 1960s.[22] Many developing countries became dependent on it, and people's dietary patterns also changed to favor wheat products (Friedmann 1982).[23]

Another key pillar in food security measures in the post–World War II era was the Green Revolution. Mainstream development communities may proclaim it as a triumph of modern science that doubled food supplies in twenty-five years (see, e.g., Rosegrant and Hazell 1999), but critics have pointed out negative ecological impacts from the intense use of agrochemicals as well as the widening of social inequality as the input-intensive Green Revolution tended to add debt for farmers (Shiva 1991). Displaced peasants constituted a labor reserve for industrial sectors that were privileged over agriculture (McMichael 2005). While the Green

Revolution decreased dependence on US wheat, it increased dependence on industrial inputs such as chemical fertilizers (Friedmann 2005, 243).[24] The Green Revolution has also been interpreted as an American Cold War strategy to contain Communism by increasing food production (Perkins 1997) while simultaneously promoting trade and investment for the Western private sector (Brooks 2010; Cullather 2004; Kloppenburg 2004).[25]

In this book I draw on studies that have critically analyzed interventions to combat food insecurity in developing countries, and I situate the micronutrient turn in the contested narratives of antihunger, antimalnutrition projects that often resulted in utopian technical fixes (Belasco 2006). In particular, agrofood studies' sensitivity to historical and geopolitical contexts is helpful in understanding interventions into Third World food problems. For instance, food regime theorists have created a thoughtful framework for understanding how postwar food aid acted as a stabilizer for the US agricultural sector by providing an outlet for surplus wheat. The micronutrient turn ought to be analyzed against the backdrop of neoliberalization, legitimated through WTO rules and related free trade agreements. McMichael (2005) identifies this as the "corporate food regime," whose critical component is the privileging of corporate power over the state. It is against a background of such historicized and political understandings of discourses on food insecurity that I analyze the rise of hidden hunger, fortification, and biofortification.

Fortification and biofortification have been analyzed in agrofood studies, but often separately as part of a social and cultural fascination with vitamins, on the one hand, and with agricultural biotechnology, on the other (see, e.g., Levenstein 1993 and Brooks 2005). I believe that their importance in the developing world cannot be understood adequately except as a part of the hidden hunger discourse that became prominent under neoliberalism, which "privatized" food security (McMichael 2005, 279).[26] McMichael observes that global trade liberalization and broad neoliberalization reframed the issue of food security as a matter of market relations. The WTO's Agreement on Agriculture in 1995 epitomized the new belief that hunger should be addressed not by national self-sufficiency but by well-functioning global trade.[27] The agreement formally rejected the right to national self-sufficiency by imposing minimum import rules and institutionalized the belief in trade as the best mechanism to provide cheap food.[28] Developing countries are to concentrate on exporting commodities where they have a "comparative advantage" and importing "cheap" commodities for their own consumption.[29] McMichael concludes that "consistent with the neo-classical agenda, 'food security' came to be redefined, and institutionalized, in the WTO as an inter-national market-relation" (276).[30]

This paradigmatic shift in the concept of food security has also manifested itself in nutritional terms. The micronutrient turn in the 1990s was propelled by, and simultaneously further justified, the thought that the market (and trade) under-pins food security. Rather than question why the poor in developing countries could not produce and eat nutritious food, solutions to hidden hunger, or micro-nutrient deficiencies, became synonymous with the consumption of nutrient-enriched products offered by the market. This involves a process of abstraction similar to the one that McMichael identifies with the making of "world agricul-ture" (270). Echoing the abstraction of agriculture from its social and ecological contexts, food was reduced to being a vehicle for nutrients. This is where nutrition-ism exerts a powerful yet understudied role in food insecurity discourses. Nutri-tionism naturalizes the logic that the solution to malnutrition is to add nutrients via fortification and biofortification, a supposedly cost-effective and non-market distorting solution that capitalizes on the know-how of agrofood businesses. I ana-lyze this facet of privatized food security, not as a simple manipulation by powerful corporations, but as interlinked relationships among neoliberalization, scientiza-tion, and gendered understandings of body and food in the global South.

This book is also informed by science and technology studies, locating nutri-tionism as an instance of what Foucault called "biopower."[31] Foucault observed a critical shift from sovereign power over life and death to biopower over the welfare of the population. This biopower promotes "the management of life in the name of the well-being of the population as a vital order and of each of its living subjects" (Rose 2007, 52) and is intimately bound up with the rise of modern sciences. The Third World's food insecurity exemplifies the need for the "management of life," and Foucault's work is useful in analyzing the processes involved in governing the Third World through food insecurity. Hence I focus on the role of scientific and technical expertise. Drawing on Foucault's concept of problematization, I analyze how the power of science, at the very basic level, socially and culturally creates the Third World food problem.

Contrary to conventional understandings, science's role is not only to provide tools to diagnose and rectify problems. In a profound way, science, in a complex relationship with other institutions, often creates the problem itself. This is what Foucault called "problematization," a situation in which there is a "development of a given into a question" and the "transformation of a group of obstacles and difficulties into problems to which the diverse solutions will attempt to pro-duce a response" (Foucault and Rabinow 1984, 388). With this concept, Fou-cault made explicit science's power in achieving "a modal change from seeing a situation not only as 'a given' but as 'a question'" (Rabinow 2003, 131) and in making "something into an object of knowledge" (Deacon 2000). Problematiza-tion keys our attention to the dynamic relationship between reality and scientific knowledge. Food problems do not arise automatically from "reality." Although there is a material reality that is undeniable, there are many ways to slice reality.

The emergence of the varying definitions of the "food problem" in the past several decades attests to such representational politics. The naming of the problem is significant because it creates a space for intervention. To use Foucault's terms, what is to be known ("effects of verdiction") is intimately tied to what is to be done ("effects of jurisdiction") (1991, 75). Once something is couched as a problem, interventions seem natural and expected, causing less opposition and resistance. "Problems" even invoke ethical obligations for intervention in the name of a specified target population.[32] Hence, to think about the concept of problematization is not to dwell on semantics but to consider the openings it enables for interventions with real material consequences.

This analytic move invites an exploration of the apparatus of problematization. The apparatus describes the historical processes of creating an object for knowledge; such processes include discourses, institutions, regulations, policies, and scientific writings, among others. What kind of apparatus enabled a particular representation of the food problem at a particular historical juncture? This book's narratives unpack the apparatus of food insecurity policies—the social, economic, and scientific institutions that control and manage the representation of food insecurity at a given time.

The concept of problematization will seem excessively abstract, if you think that we know exactly what the problems of the Third World poor are. Do they have enough food? Are people malnourished? What do the hungry in developing countries need? Indeed, for something like nutrition, it may seem that we should know exactly what the problem is. If nutritional problems are seen as located in the realm of hard science, and not as a social problem, then nutritional science should provide definitions unproblematically. Even social scientists who point out multiple layers of human "needs" and culturally constructed understandings of social problems (e.g., Maslow 1943) tend to exempt nutritional issues from such social understandings, and are willing to delegate authority on the subject to nutritional scientists (Douglas et al. 1998).[33] Yet contrary to the public face of nutritional science, even nutritional scientists do not have absolute certainty about "what to eat," to borrow the title of a popular book by Marion Nestle, in which she confesses that "like any kind of science, nutritional science is more a matter of probabilities than of absolutes and is, therefore, subject to interpretation. Interpretation, in turn, depends on point of view" (2002, 28).

Such candid remarks by nutritional experts are rare. Instead, scientists and experts are often in a privileged position to define the problem. Conventional demarcations between science and nonscience are a powerful obstacle to nonexperts challenging diagnoses by experts (Fraser 1989; Haney 2002).[34] It is even more difficult when the problem is said to be beyond the direct perception and recognition of lay people. As the common nickname for micronutrient deficiency, *hidden* hunger, suggests, it purports to be invisible to the lay person's eye or even to hungry people themselves. If the deleterious effects of hunger are invisible and knowable only by experts, by way of scientific measurement, people lose the foundation on which to ground their experience and the possibility of critiquing official interventions. When problems are supposedly unrecognizable without scientific expertise, the contestation between expertise and experience is even more asymmetrical. There is an urgent need to scrutinize what kinds of problems are constructed and promulgated by experts.

Criticism of the scientization of food insecurity is not to deny various contributions of science. Instead, my point—and this is informed by the growing literature on the relationship between science and democracy—is the need to explore the tension between democracy and scientific expertise (Callon, Lascoumes, and Barthe 2009).[35] Some might argue that science is inherently undemocratic (Brooks and Johnson 1991; Perhac 1996). But for issues like hunger and malnutrition that are complex—historically rooted and locally specific while simultaneously involving global factors, and encompassing social, natural, and human sciences—the need for democratic discussion is compelling. Food regime theorists have pointed out that we need to be aware of the historic specificity of our time. They have noted that the current food regime is increasingly controlled by the private sector, unlike the nation-state–based regime of the 1940s through 1970s. The growing power of the private sector is also reflected in technical and scientific fields. The private sector is now a major source of financial resources and intellectual property in scientific research, and even research by public research institutions is often done in "partnership" with corporations and/or dependent on information and materials that are the property of the private sector (Brooks 2005).[36] How could private corporations have come to dominate the research agenda and the way the results are disseminated and used? Private firms are not accountable to citizens in the same way that public institutions are. The scientization of food insecurity, particularly in the context of growing corporate power in science and technology, demands that we question its implications for democracy and governance.[37]

Quantity vs. Quality

The micronutrient turn in food policy was often portrayed as a welcome change of attention to quality in contrast to the previous emphasis on quantity of food. From hunger to hidden hunger, from quantity to quality, the micronutrient turn in international policy has been portrayed as a radical departure from the earlier focus on caloric intake, famine, and agricultural modernization. Indeed, the productivist paradigm best exemplified by the Green Revolution has been heavily criticized for its social, ecological, and nutritional consequences (see, e.g., Shiva 1991). My criticism of the nutritionism that shaped the micronutrient turn in food policies could easily be interpreted as support for the productivist approach or a blanket rejection of attention to "quality." However, I argue that productivism and quality-based approaches are two manifestations of scientized views of food insecurity that surfaced at different historic moments.[38]

Despite the rise in interest in micronutrients and the rhetoric of quality, the productivist approach is far from extinct. Since the mid-2000s, and particularly after the food crisis of 2007–8 when food prices soared and food riots erupted in many countries, there has been a renewed emphasis on the productivist approach. For instance, the World Bank's *World Development Report 2008* focused on agriculture for the first time in twenty years and pledged to increase funding for agriculture (World Bank 2007). The Bill and Melinda Gates Foundation, which has become a major player in the area of global health, similarly started to channel substantial resources to agriculture in the mid-2000s (Bill and Melinda Gates Foundation 2011).[39] In addition, in the midst of the food crisis, micronutrients seemed to take a backseat. In Indonesia, mandatory fortification of wheat flour was stopped due to industry lobbying over the skyrocketing price of wheat flour. Yet biofortification and fortification today remain on the agenda of organizations such as the World Food Programme and the Consultative Group on International Agricultural Research (CGIAR).[40] Hence, the notion of a paradigm shift from the productivist to the nutritionist is insufficient: they coexist.

Even though productivist and "quality"-oriented projects might look mutually exclusive, the criticisms of nutritionism that I summarized earlier actually capture many of the troublesome aspects of current productivist policies as well. It is instructive to examine food policy discourse after the 2007–8 food crisis. Even with a renewed emphasis on productivist agricultural programs, the discourse shares some of the key aspects of nutritionism. First, just as nutritional fixes have emphasized technical interventions over social and political ones, so many of the current productivist proposals focus on the intensification of agriculture through technological packages of high-yielding varieties, fertilizer, irrigation, and biotechnology. This trend is epitomized in calls for a second Green Revolution and a so-called gene revolution that portray the future of agriculture in the global South as lying in modernizing technologies. One such example is the Rockefeller/Gates Alliance for a Green Revolution in Africa that recently started with more than $150 million in funding (McMichael 2009).

Second, the food crisis was accompanied by much discussion about causes. But the debate was primarily technical in nature, with little attention to broader structural issues in food systems. For instance, policymakers fiercely debated the degree to which various factors contributed to the calamity, whether biofuel production, export restrictions, productivity slowdown, rising oil prices that affected prices of agricultural inputs, declines in grain stocks, or the increasing demand for grains from developing countries was the culprit (see, e.g., Headey and Fan 2008).[41] These are doubtless important factors, but what this discussion omits are larger problems beyond the immediate supply, demand, and trade of food—such as the coupling of global financial markets and food markets, the decline of smallholder agriculture in developing countries through neoliberal policies, the "corporate food regime" that concentrates and centralizes the power of agribusiness through government policies, and "food empires" that have increased the overall vulnerability of the food market to external shocks (Ghosh 2010; Lang 2010; McMichael 2009; van der Ploeg 2010).

The micronutrient turn in the 1990s cannot be understood without acknowledging the impact of neoliberalism, and neoliberalism profoundly shapes the current productivist approach as well. While Green Revolution programs increasingly at first fell out of favor due to neoliberalism, agricultural projects are now urged to seek the power of the market and to tap private sector resources. For instance, the *World Development Report 2008* urges that agricultural development be led by "private entrepreneurs in extensive value chains linking producers to consumers and including many entrepreneurial small holders supported by their organizations" (World Bank 2008, quoted in McMichael 2009, 236). The Consultative Group on International Agricultural Research has adopted new organizational mechanisms in order to serve as the "broker" for private and public research institutions and to increase public-private partnerships (IFPRI 2005, cited in Brooks 2010).[42]

Finally, the marginalization of the poor and undemocratic food policies are not addressed in the current productivist approach. In analyzing the World Bank's espousal of "New Agriculture," Philip McMichael is critical of the way it still considers small farms in Third World countries as inefficient and in need of "development" and modernization. How women farmers in the developing world, who tend to be subsistence farmers, might be affected by New Agriculture is not analyzed in the World Bank's approach. Similarly, in their critique of the Alliance for a Green Revolution in Africa, Holt-Gimenez, Altieri, and Rosset (2006) point out that AGRA's advocates have only "consulted with the world's largest seed and fertilizer companies, with big philanthropy, and with multilateral development agencies, but have yet to let peasant farmer organizations give their views on the kind of agricultural development they believe will most benefit them" (8). Reporting from Rwanda, which has collaborated with AGRA to increase agricultural yields, Miltz (2011) writes that the program frequently coerces peasants to conform and that it "is not a consensus-driven process; there is no attempt to consider the needs and opinions of the main people affected.... [The] Rwandan government, led by its charismatic president Paul Kagame, opted to rule the ag sector with a heavy hand. Put bluntly, it is frog-marching the country down a particular rural development road, with little allowance for debate or criticism."[43]

Rather than quantity and quality in food policies being diametrically opposed issues, I believe that the policy discourse changes its emphasis depending on various social, political, economic, cultural, and scientific factors. My emphasis on nutritionism as the driving force behind the micronutrient turn helps us to look beyond superficial differences and understand deeper problems. Both quantity- and quality-based approaches are scientized and undemocratic, privileging a select group of experts over the poor and the hungry.[44]

Friedmann (2005) observes that when the food crisis of 1974 took place, redistribution to address growing inequality between the rich and the poor (domestically and globally) could have been a response. Instead, the 1980s and 1990s saw the "triumph of neoliberal polices centered on trade and finance," and "advocates of free trade pinned hopes on technological change, now including genetic technologies" (248). The trajectory of the 2007–8 food crisis echoes this historical pattern of depoliticization, recasting a political problem of food insecurity as a technical problem. Social movements had been mobilizing alternative ways of addressing food insecurity, notably that of food sovereignty as proposed by the global peasant movement Via Campesina. Mainstream policies responded to the food crisis with little engagement with these movements, refusing to address global inequality, de-peasantization, the dismantling of social protections, export-oriented agriculture displacing subsistence farming, or environmental degradation. Rather, they opted again to emphasize technical, market-based solutions.

Attention to "quality" of diet could have presented a profound criticism of the dominant policies that promoted "modern" agriculture with high dependence on agrochemicals, reduced the diversity of crops planted by farming communities, and increased dependence on processed and imported food products in the global South. Instead, the quality discourse was watered down to technical matters, providing further opportunities to avoid structural issues.[45] Both the mainstream "quality" and productivist approaches end up reducing food insecurity to the need for technical-scientific interventions, offering scientized descriptions (definitions) and prescriptions (solutions). In this sense, the quality-oriented policies of the 1990s did not open a radical new frontier in international development. Defined and acted on exclusively by experts, Third World food insecurity was still problematized in a microscopic, reductionist manner that did not challenge existing power.

The Plan of the Book

I begin this book by examining the rise of micronutrients and fortification and biofortification in the 1990s on an international level, as it appeared within the community of experts in the business assisting developing countries. In chapter 2, I situate the micronutrient turn in a longer history of changing discourses of malnutrition and the "food problem." I trace multiple representations of Third World food issues with the concept of "charismatic nutrients." I also examine a series of nutritional fixes, from high-protein cookies in the 1960s and vitamin A capsules in the 1980s to fortified and biofortified food products in the 1990s. The description of the food problem, presented as a straightforward product of science and technical calculations, is actually historically contingent.

In chapter 3, I provide more details about the micronutrient turn by investigating international commitments and agreements regarding the eradication of micronutrient deficiencies in the 1990s. Untangling the micronutrient network that has evolved shows several important factors that have led to a particular shaping of the "problem" and the "solution." I highlight the critical role played by the World Bank and other multilateral lending organizations. Fortification in particular received much advocacy from these organizations, and I suggest that its high resonance with the neoliberal ideology is an important factor.

Discursive analysis is often criticized for being abstract and universalistic. While the micronutrient turn was a global discursive change, its impacts were locally varied and discontinuous. In chapter 4, I analyze historical changes in food and nutrition policy in Indonesia. Has Indonesia seen a micronutrient turn? If so, should we see merely the diffusion of an international norm, or can we find specific reasons behind it? I answer these questions by examining the dynamic relationship between international and local actors.

Chapters 5, 6, and 7 are devoted to individual commodity studies in Indonesia. Each commodity exemplifies and manifests the concrete operation of one of three principal micronutrient strategies: (1) mandatory fortification, (2) voluntary fortification, and (3) biofortification. Chapter 5 looks at wheat flour fortification, which became mandatory in Indonesia in the late 1990s. Chapter 6 examines baby food as a case of voluntary fortification. Chapter 7 examines Golden Rice, the most famous biofortified crop. All three commodities—wheat flour, baby food, and biofortified rice—were meant to solve the food problem for Indonesians. How did they end up being accepted as solutions? What unites these commodity analyses is the construction of the food problem under nutritionism.

Having achieved strong economic growth in the post–World War II era, Indonesia might easily be taken as best evidence of the naturalized understanding of the micronutrient turn: that the interest in micronutrients increased because the country had solved the traditional quantity problem and now was taking on micronutrient deficiencies with the aid of more advanced science. Such an apolitical understanding of the micronutrient turn fails to account for its complex configuration of science, policy, and markets.

I do not claim that the Indonesian case is generalizable to all other so-called developing countries. The dynamics in Indonesia differ from those in countries that are struck by famine. The specificities of food industry; government; nutritional, health, and agricultural research institutions; and people's dietary patterns also influence the rise and fall of interest in charismatic nutrients and nutritional fixes in different countries. Nonetheless, the rise of "smart food" in Indonesia has many parallels in other parts of the world—for example, fortified yogurt advanced by Danone and the Grameen Bank, Plumpy'nut that was at first used for emergency intervention in Africa but now marketed as "malnutrition prevention," and Horlicks drinks and instant noodles ("taller, stronger, sharper") sold by Glaxo Smith Kline in South Asia.[46] These cases suggest that Indonesia is not alone in witnessing the growing influence of nutritionism and the concomitant rise of smart food. The Indonesian cases help us identify important variables and concepts that can be of use in analyses in other countries.

In sum, the book unpacks the rise of smart food as an antimalnutrition, antihunger strategy in the Third World. The strategy entails a growing scientific gaze on food insecurity, extended and reconstituted through the network of global capital, international development programs, national governments, and scientific experts. At the core is the power of knowledge to create a certain lens through which we see food, the body, and health. The growing influence of nutritionism means that quality is translated into a narrower set of nutritional parameters, food insecurity into a problem of deficiency of some nutrients. The solution is conceived as filling the nutrient gap by channeling more nutrients into bodies by means of biofortified crops and fortified food products. In this way, nutritionism constricts the boundaries by describing a reality in which only certain kinds of responses become imaginable. These smart foods appeared as a radical new way to replace the productivist program that championed "technological packages" for modernizing agriculture. However, they actually preserve the tenets, keeping the existing power structure of international development and global capital intact. When hunger is conceived as a technical problem to be solved by experts, the poor—particularly impoverished women—are still marginalized. Nutritional interventions are made in the name of the well-being of these women and their children, but poor women are not seen as active agents who define and address their own food insecurity. Instead, it is the experts—from the scientific community and increasingly from the private sector—who are seen as the benevolent "doers" who fight global hunger and malnutrition.

2

CHARISMATIC NUTRIENTS

The focus of nutrition interventions evolved from control of protein deficiency, followed by concern about protein-energy deficiency, to the prevention and treatment of micronutrient deficiencies.

—Lindsay H. Allen, nutritional scientist, 2003

Micronutrients are not the first instance of the scientization of food security through nutritionism. Definitions of food problems have changed, and this is not necessarily because of changing circumstances in food production and consumption and scientific advancements. This brings us back to the concept of problematization, which shifts the emphasis from "the truth" to "the representation of truth." How a social problem is defined and presented is historically contingent. The apparatus of problematization produces certain visibilities and invisibilities, creating "schemes of possible, observable, measurable, classifiable objects" (Foucault 1981, 55). By seeing it as historically specific problematization, we can understand any representation of food problems, priorities, and needs as a construct that exists within a particular political context.

Scientization of food insecurity through nutritionism can be viewed as successive eras of different *charismatic nutrients*. These nutrients come to command center stage in international food and nutritional politics when their suboptimal intake defines the nature of the food problem in developing countries. Before micronutrients, there were other charismatic nutrients, and in this chapter, I discuss protein in the 1950s and 1960s and vitamin A in the 1980s.[1] I use the word "charismatic" nutrients after Max Weber's classic theory of charismatic authority, that is, a leader who exudes authority beyond normal expectations. Weber's use of the term is helpful, as he astutely noticed that charisma was a social status rather than a personal quality, hence irreducible to a divine endowment (Weber 1978, 241). Following Weber, I argue that the charisma of nutrients cannot be fully captured by their "scientific" values, but rather, depends on sociopolitical networks built around them. In other words, vitamin A's charismatic status in the 1980s, for instance, cannot be fully explained by its physiological potency, but only by disentangling the social relations that formed around it at a particular historical juncture. Another important insight from Weber is that charisma is not a stable form of authority but, rather, "quicksilver, unstable" (Smith 2007; see

19

also, Weber 1978, 242, 1141). Different phases of international food politics have problematized different charismatic nutrients, which became the focal points for international and domestic food interventions. Changes in the focus of food policies cannot be considered simply as a result of "progress" in science. Rather, I argue that the ebb and flow of the locus of the "food problem" with different nutrients attests to the transient nature of the "charisma" of nutrients.[2]

My application of the concept of charisma to the global politics of food is also inspired by research on global environmental politics. In environmental movements, "charismatic megafauna," such as pandas or elephants, play an important symbolic role. Without such an icon for an NGO's logo or magazine cover, the narrative of the impending "global ecological crisis" and the call for conservation would be less powerful. Like these animals, *charismatic nutrients* help to focus global food policies by embodying the problem, capturing both the public and development experts with a compelling tangibility that stands for complex problems.

This chapter demonstrates how charismatic nutrients have been a product of complex social relations and highlights the crucial roles of sponsors and a sponsoring discipline (nutritional science), with accompanying "facts" and "fixes." Four dimensions of charismatic nutrients are important. First, for any charismatic nutrient, there are organizations and experts who try to sell its potency, benefits, and morality to other actors and organizations. The charisma of particular nutrients increases in the hands of these capable sponsors in international organizations and scientific communities. Second, charismatic nutrients are accompanied by various nutritional fixes. That their power can be delivered to the poor in a relatively simple form to solve a complex problem enhances their lure. Third, the boundary making and institutional building of nutritional science, and more specifically international nutrition as a discipline, has had a major impact on charismatic nutrients. Set in the broader field of international development, charismatic nutrients make tangible the legitimacy of "nutritional science" as a relevant field of research and action for the global endeavor of improving the world's food situation. By embodying the essence of food insecurity, charismatic nutrients have become the icon of Third World dystopia and nutritional science as the essential science for solving the problem. Fourth, charismatic nutrients have gendered implications. The charisma of nutrients has been crucially linked with children as their beneficiaries. Such an attitude, in policy and scientific discourses, often casts women as reproductive beings in relation to their (actual and future) children. I will explore the implications of this mother-child dyad viewpoint.

On the surface, charismatic nutrients successfully symbolize the reality of the food problem. As with the use of charismatic fauna in global environmental politics, however, the use of any charismatic nutrient incompletely captures the politics of the problem it intends to address. It is incomplete because it reduces a complex social, cultural, and historical situation to a problem with a single focus ("Save the whales!" "Give them vitamin A pills!"). While the image of "the problem" as a matter of missing nutrients can be institutionally and culturally appealing because of its simplicity, that very simplicity belies the much messier reality of the global food problem. Fallen outside the aura of charisma are less glamorous actors and multiple layers of problems that pervade the world food system.

The Protein Fiasco

No milk, no meat, and no eggs. Shriveled vegetables and rice porridge. Historically, the lack of protein was understood to be the defining feature of the Third World food problem: "As many as one-third of the children are estimated to suffer from protein malnutrition, and it is feared that if this situation continues, the physical, economic, and social development of the future generation may become completely arrested," a United Nations advisory committee noted in 1968 (quoted in Carpenter 1994, 162). Scholars thought that not only the insufficient amount of food but insufficient protein explained the ill-health of people in developing countries. Protein emerged as a charismatic nutrient by the 1960s, becoming the focal point of various international programs that were committed to addressing the "protein gap" (Carpenter 1994, 161; Ruxin 1996). For instance, in 1952, the newly established Joint FAO/WHO Expert Committee on Nutrition agreed that "protein malnutrition" was a "problem of fundamental importance throughout the world" (Joint FAO/WHO Expert Committee on Nutrition 1952, 4). In 1955, WHO's Nutrition Section similarly concluded that "kwashiorkor is without doubt the most important nutritional public health problem of the present time" (quoted in Ruxin 1996, 72). The urgency with which protein deficiency was viewed as a global problem was clear from reports commissioned by the UN, such as *Action to Avert the Impending Protein Crisis* (UN Advisory Committee 1968) and *Strategy Statement on Action to Avert the Protein Crisis in the Developing Countries* (UN Panel of Experts 1971).

But how did protein come to be the marker of the Third World food problem? One important factor was that a disease called kwashiorkor came to embody protein deficiency. Kwashiorkor is a disease that is observed in developing countries plagued with famine and political unrest, and its symptoms include changes in skin, diarrhea, fatigue, hair loss, infection, failure to grow, protruded belly, and edema. With its striking visual signs, including large bellies, edema, and the depigmented skin of infants, kwashiorkor powerfully symbolized the misery and poverty of the Third World, which further cemented experts' fascination with protein since there was an emerging consensus that the disease was caused by protein deficiency.

Yet the charisma of protein was not built overnight, and historical and cultural heritage helped its ascendance. Protein already had a special status in Western culture, perhaps accounting for the initial demarcation of it as the signifier of what was missing from the non-West's diet (Cannon 2002). Since the nineteenth century, protein had been thought of as the principal nutrient that builds organisms. The power of protein had been popularized since the mid-nineteenth century, thanks in particular to the German chemist Justus von Liebig. He argued that protein was the "master nutrition" of living organisms and the central building block of the body and propagated the idea of the importance of protein among the general public, even selling a concentrated protein from beef as Liebig's Extract (Semba 2001). Nutritionists after him were similarly fascinated by protein, one of their most heated scholarly preoccupations being determining the exact quantity of protein needed to nourish the human body (Cannon 2002).

In addition, protein enjoyed an iconic status in the field of international nutrition. Protein figured critically in the interpretation of the social problems of colonies in the days of empire building, and in the conceptualization of indigenous "inferior food." Many influential nutritional studies of the colonies were explicitly founded on the notion of protein as the primary human nutrient. For instance, a seminal study by two British scientists, John Gilks and John Boyd Orr (the latter became the first director-general of FAO in 1945), on tribal health and diets in Africa compared the diets and health of the Kikuyu and the Maasai and linked divergent health outcomes with their respective dietary patterns. They argued that the "inferior" physique and health of the Kikuyu people was attributable to the lack of protein in their vegetarian diet. Their article in the medical journal the *Lancet* concluded that it was protein that critically determined the divergent health outcomes (Gilks and Orr 1927). Subsequently, the notion that lack of protein characterized the inferior diet of the colonies began to take hold and several nutrition improvement programs that focused on protein were conducted (Worboys 1988; Brantley 1997).

Juxtaposed with colonial racist assumptions, protein even provided a scientized explanation for the West's perceived superiority to the Orient. A widely used medical textbook by J. S. McLester of the University of Alabama (1939), for example, argued that "the prowess and achievements of our early Anglo-Saxon ancestors have been attributed in part to the energy-giving effects of the meat which they consumed in liberal quantities" and "if man would enjoy sustained vigor and would experience his normal expectancy, as well as contribute to the improvement of his race, he must eat a liberal quantity of good protein" (77).[3]

These historical and cultural forces were at play when protein achieved its stardom in international food policies in the post–World War II period.

While these accumulating historical forces help explain the growing social appeal of protein in this period, it is also interesting to ponder why protein did not immediately come to occupy the central place in international development immediately after the war and only in the 1960s. We can consider several reasons. First, the Third World food problem did not emerge as the problematique for the international community until the situation in Europe saw a significant improvement. The "food problem" did not have the obvious spatial connotation that it does today. In fact, the devastation in Europe preoccupied international organizations, and the bulk of international aid was directed toward Europe (Ruxin 1996). At that time, the "world food problem" was considered a European problem.

In considering the relatively slow arrival of protein as representing the food problem, the gendered history of protein in international nutrition has to be considered as well. The history of kwashiorkor echoes the history of nutritional science, in which female academics confronted the hostility of male colleagues (Apple 1996). Kwashiorkor was first reported by Cicely Williams, a female British doctor, in the 1930s (Williams 1935). She was the first to use the term "kwashiorkor" and suggested the protein deficiency as the cause (Carpenter 1994). Although her work was pioneering, other experts ignored it. Her article on it in the *Archives of the Diseases of Childhood* received only one response, a critique by another doctor who charged Williams with misdiagnosing a form of pellagra.[4] It took decades before other scientists began to build directly on Williams's work. Fifty years after its initial publication, the journal finally noted that her article was the most important article in its history and reprinted it (142).

Protein needed more powerful "sponsors" than a female doctor working at the periphery to give it a boost. Protein's status greatly improved when key nutritional scientists active in international organizations, such as John Conrad Waterlow and Nevin Scrimshaw, advocated a protein-based understanding of food problems (Carpenter 1994; Ruxin 1996). Both Waterlow and Scrimshaw were pioneers in international nutrition, having worked in developing countries and with development agencies since the 1950s. Waterlow worked for the British Colonial Office in Jamaica, and taught human nutrition at the University of the West Indies and later at the London School of Hygiene and Tropical Medicine. Nevin Scrimshaw was the head of the Department of Nutrition and Food Science at Massachusetts Institute of Technology (MIT) and founder of the Institute of Nutrition of Central America and Panama, a part of the Pan American Health Organization. As is clear in comments that Scrimshaw made in the *New York Times,* that "not only do many people have too little food but what they do have contains little or no proteins" (Nagle 1976), key nutritional scientists joined force to lobby for the importance of protein.

The sociopolitical network supporting protein's charismatic image further expanded when several UN organizations, including WHO, FAO, and UNICEF, established the Protein Advisory Group in 1955. As its name amply demonstrates, the PAG was primarily concerned with the protein deficiency problem, and was established to advise on the issue and to spur international collaboration on engineering protein-rich food. Scrimshaw became its chairman, and PAG started publishing the *PAG Bulletin* in October 1957 to disseminate information on research and development of protein-rich foods around the world (UN ACC/ SCN 1978).

By narrowing down the food problem to protein deficiency, the international development community was able to move swiftly from defining the problem to engineering the solution. Milk was one of the first products that international organizations identified as a solution to the protein deficiency problem. UNICEF already had experience with a dairy industry assistance programs in Europe, and the promotion of milk seemed like an ideal program for the Third World context as well. On the recommendation of the Joint FAO/WHO Expert Committee on Nutrition, emergency food aid started to include skim milk distribution. But experts wanted more than milk. They started to seek the "ideal" protein-rich food that could deliver an optimal amount of protein in the most efficient manner. The committee, for instance, identified six products as the ideal raw materials to be engineered into a super protein food.[5] Various organizations started to pursue the creation of a super protein food.

Nutritional fixes engineered to tackle the "protein gap" had an impressive product lineup. Historian of nutritional science Kenneth Carpenter profiled (1994) various projects by experts seeking to engineer protein-rich food in this period (summarized in table 2.1). For instance, the UN funded a Chilean government manufacturing plant to produce fish flour. UNICEF helped the Nigerian government purchase and distribute a commercial baby food called Arlac that was made of peanuts. UN agencies provided funding for the Indonesian government to develop and sell soy milk. The Institute for Nutrition in Central America and Panama developed a flour mixture from cottonseed and encouraged governments to market it.

Universities and governments from the developed world were also eager to participate in this international mission. MIT, which had one of the leading international nutrition programs in the United States, started a project to develop protein-rich supplements and protein concentrate from fish (Carpenter 1994). The US government experimented with fish flour, seaweed, and petroleum derivatives (Carpenter 1994; Belair 1965) as illustrated in the following excerpt from the *New York Times* (November 25, 1968):

TABLE 2.1 Examples of protein-rich food projects

RAW MATERIAL	YEAR	AGENCY	PRODUCT	RESULT
fish	1958	Chilean government with UN funding	fish flour	The government decided to drop the program.
fish	1961	US Bureau of Commercial Fisheries	fish meal	The raw material became expensive, and the project was dropped.
peanuts	1963	UNICEF with Nigerian government	weaning food	The marketing did not go well, and UNICEF stopped support.
soy	1957	UN and Indonesian government	soy milk	UN aid was conditional on using and distributing the product for free to the poor, but it was only sold to the well-off. The UN cancelled its support after ten years.
cottonseed	1961	Institute for Nutrition in Central America and Panama (INCAP)	flour mixture weaning food	It was only commercially sold, and the price was unaffordable for the poor.

Source: Carpenter 1994.

Members of the United Nations Economic and Social Council interrupted their proceedings the other day to munch approvingly on chocolate chip cookies provided by the American delegate, Arthur Goldschmidt. The cookies were made from fish flour. Eighty-four Michigan farmers and their wives at a dairymen's meeting last year toasted the cow with big glasses of what they thought was good rich milk. Only two suspected they were really drinking an imitation made from palm oil, corn syrup and seaweed extract. In Bihar, last year's near-famine state in eastern India, peasants are eating chapattis, the traditional unleavened bread, baked in the traditional way on an ungreased griddle. The chapattis taste the same, but they have been prepared from American Food for Freedom flour which has been fortified with amino acids, derived from waste carbohydrates or petroleum, to provide the protein that is lacking in the average Indian diet. (Brown 1968)

Their willingness to try anything to engineer the magical protein-rich food is almost humorous, yet these nutritional fixes commanded serious commitment from diverse international organizations and governments. The possibility of creating super protein products fascinated bureaucrats and scientists in the field of international development with their modernist promise of providing an uncomplicated solution. Such promise further enhanced the charisma of protein as the key signifier of global food insecurity.

Outcomes of the Protein Fiasco

According to Carpenter (1994), these high-protein nutritional fixes ultimately failed to improve the nutritional status of the Third World poor, however. The fish meal project in the United States failed, for instance, because the stock of the fish species being used as raw material collapsed. The projects of the Institute for Nutrition in Central America and Panama did not reach the target population. Indonesian soy milk was never distributed to the poor as had been promised by the government. By the early 1970s, international organizations such as UNICEF had started to realize the widespread failure of protein solutions (Ruxin 1996).

Further dampening enthusiasm for the protein fixes was the realization by nutritional scientists that protein deficiency rarely occurred independently of caloric deficiency (Levinson and McLachlan 1999). The definition of kwashiorkor as a protein-deficiency disease became suspect, although its etiology as a protein-deficiency disease was the original reason for scientists' heightened interest in protein. A growing number of studies found that kwashiorkor could be treated without high-protein food.[6]

Now the protein-gap model was in doubt. Some experts increasingly believed that the focus on protein had led to the gross neglect of the problems of inadequate calories and insufficient quantity of food. Worrying that nutritional science's misplaced priorities caused a neglect of the more prevalent condition of marasmus, which is a form of protein-energy malnutrition, some scientists, including Donald McLaren at the American University of Beirut (1966), began to criticize the protein-gap model. Calling contemporary scientists' enthusiasm "the great protein fiasco," McLaren lamented that "millions of dollars and years of effort that have gone into developing these [high protein] foods would have been better spent on efforts to preserve the practice of breast feeding…being abandoned everywhere" (McLaren 1966, quoted in Carpenter 1994, 184). Even J. C. Waterlow, who had been a staunch supporter of the protein-gap model, had to admit that the idea of the protein gap, although it had fuelled a tremendous amount of international effort to seek protein-rich food, was no longer valid (Waterlow and Payne 1975; Carpenter 1994, 228).[7]

This kind of reassessment suddenly made the scientific standard for the human protein "need" uncertain and problematic. While experts had agreed that the protein requirement for a one year old child of breast milk was 2.0 g/kg of body weight per day, they had to resume the debate, eventually reducing it to 1.1 to 1.2 g/kg per day (Carpenter 1994, 184). This revision of the previously accepted scientific standard raised profound questions regarding the claims for the role of protein deficiency. The need for protein-rich food was now in great doubt, as it seemed that the availability of protein in the diet of most countries' populations was sufficient to meet requirements (Cannon 2002).

Charismatic nutrients were difficult to let go of, however. Experts did not immediately relinquish their old framework. They were in a difficult situation. They did not want to reject the importance of protein altogether, as they had invested so much in it, but at the same time they had to acknowledge that something else was going on.[8] Yet Ruxin (1996) noted that the UN was still focused on protein in their actual programs for a while, and it was only in 1971 that the Joint FAO/WHO Expert Committee on Nutrition finally admitted that there had been "a tendency to overemphasize the importance of either protein or calorie deficiency alone, whereas in fact the two almost always occur together" (quoted in Ruxin 1996, 252).

The "protein fiasco" is instructive not only because it points to the degree to which a charismatic nutrient received financial and other resources to fill a supposed deficiency. While one might assume that global food insecurity came to be defined as protein deficiency in the 1950s simply as a result of scientific advances, the charisma of protein also involved social, cultural, and political factors. Employing historically powerful imageries of protein in human nutrition and colonial misery, it built an impressive network of nutrition experts and international organizations. The striking visual image of children with kwashiorkor was another powerful symbol of the need for protein. The imagery of starving children added a moral persuasion to protein.

The story of protein also points to the gendered implication of charismatic nutrients. The icon of starving children in developing countries was inevitably accompanied by the scrutiny of mothering practices. Concern for the welfare of children in the developing world often resulted in blaming mothers as ignorant, but it rarely translated into commitment to mothers' welfare and improvement in their overall living conditions. While the attention to children's protein status brought mothers into the expert discussion, experts tended to see the primary importance of mothers as a pathway to their children's bodies. I will come back to this point later in the chapter.[9]

Vitamin A: The Magic Bullet

Protein was not the only particular nutrient that became a focal point in the discourses surrounding the Third World food problem. In fact, the general interest in micronutrients in the 1990s was presaged by a strong interest in vitamin A deficiency in the 1980s. Before the term "micronutrient" became widely accepted as a label for a host of nutrients, vitamin A had single-handedly become a development buzzword. "There are very few wonder drugs in the world, but vitamin A may be one of them," the *Washington Post* noted on November 7, 1994 (Brown 1994). International organizations promoted vitamin A deficiency as the most important disease to be tackled. For instance, UNICEF and WHO started to recommend free distribution of vitamin A supplements in developing countries. In 1989, the US Congress decided to earmark $8 million for vitamin A supplements (Edmunds 1989). A number of governments of developing countries started to provide vitamin A capsules to children. Indonesia, for instance, accelerated its vitamin A supplement program (Shaw and Green 1996). In 1992, the UN's Administrative Committee on Coordination, Sub-Committee on Nutrition (ACC/SCN) recommended vitamin A capsules as a possible tool to reduce child mortality (Underwood 1998).

How did this new charismatic nutrient come to the fore? One catalytic event often credited with ushering in the vitamin A epoch was the publication in the *Lancet* of research done on vitamin A's impact on child mortality by Alfred Sommer at Johns Hopkins University. Sommer conducted a survey of children in Aceh, Indonesia, in conjunction with a nongovernmental organization (NGO), Helen Keller International, and the Ministry of Health of the Government of Indonesia. This research—later known as the "Aceh study"—demonstrated that giving preschoolers vitamin A supplements at six month intervals reduced their mortality by 34 percent (Sommer et al. 1986). That vitamin A deficiency caused eye disease had long been known, but this study demonstrated its effect on mortality. The study had a large sample of 29,939 children from 450 villages and was randomized, which added to its scientific credibility. The Aceh study became widely influential.

On the surface, the charisma of vitamin A might be ascribed to the novelty of Sommer's research. Yet scholars have found that the Aceh study was not the first to point out the link between vitamin A and mortality. In the 1920s, Edward Mellanby and Harry Green at the University of Sheffield in England had found that vitamin A deficiency led to increased infections in animals. They theorized that vitamin A plays a significant role in enhancing the body's resistance to infection (Mellanby and Green 1929). This theory of vitamin A as an "anti-infective" vitamin led to many studies on vitamin A as a means to reduce morbidity and mortality. More than thirty studies were conducted to determine whether vitamin A could reduce the morbidity and mortality of measles, puerperal sepsis, and other infectious diseases. Historian of nutritional science Richard Semba notes that "the public seized upon the use of vitamin A as anti-infective therapy [in the 1920s], but the value of vitamin A in reducing morbidity and mortality from infections was not more widely recognized until 50 years later" (1999, 783). So why did vitamin A become a new charismatic nutrient when it did? To answer this, we need to understand the historical context and the sociopolitical network around vitamin A. "Nutritional isolationism" is important in understanding the historical context (see Levinson 1999). In the 1970s, the field of international nutrition saw a push for what was called multisectoral nutrition planning. Articulated most clearly by Alan Berg of USAID in *The Nutrition Factor* (1973), MNP proponents pointed out that development projects had neglected nutrition while prioritizing other sectors such as agriculture and education. They argued that nutritional science should be taken more seriously in the international development sector. They also argued that the past failure of international nutritional efforts was due to the lack of cooperation from other sectors (Escobar 1995; Levinson 1999); they posited that the same mistakes would be made without a "multisectoral" approach. However, by the mid-1980s, MNP came to be seen as a failure. One discussion of MNP in the journal *Food Policy* declared the death of the approach with the provocative title *MNP: A Post-Mortem* (Field 1987). This led the nutritional sector to resolve that if other sectors did not want to collaborate with them, then they should carry out projects on their own. In this context of "nutritional isolationism," the preferred framing of the problem was in strictly nutritional terms, and vitamin A supplied a space for nutritional experts to operate in.

Additionally, just as with protein, powerful sponsors played a critical role. One important reason for vitamin A's success in the 1980s involves vested interests within organizations, including experts who considered themselves members of the "vitamin A gang." Originally called the International Vitamin A Board, the International Vitamin A Consultative Group (IVACG) provided an ideal space for networking, lobbying, and seeking international support. Founded in 1975 by USAID and international experts, IVACG was the major international arena for discussion of vitamin A-related issues.[10] Although its founding meeting in Jakarta and subsequent meetings were filled with scientific disagreements and conflicts (Underwood 2004), these disagreeing experts were nonetheless bound by their commitment to this nutrient. Indeed, the Jakarta meeting signaled the emergence of a community of experts—both scientific and policy oriented—focused on promoting the up-and-coming charismatic nutrient.

Vitamin A had a very effective spokesperson on its behalf: Alfred Sommer. Sommer, the principal researcher of the Aceh study, was central to the vitamin A gang. I spoke with Sommer in September 2004. He emphasized the scientific underpinning of vitamin A's ascendancy, but it was clear that his political and social skills were also crucial. Dressed in a dark business suit and with a practiced pitch about vitamin A's effectiveness, Sommer seemed more like a high-ranking diplomat than a researcher. Indeed, his skill in building a scientific network in support of vitamin A in academia and with politicians, international organizations, and media was crucial. He recalled his various efforts to build a vitamin A network and emphasized his expertise in dealing with controversies. In his view, being controversial was not necessarily a bad thing, as it offered opportunities to expand the vitamin A network. The Aceh study was a controversial piece of research, and it caused much debate within academia. Many scientists thought that the result was random or that there had been a flaw in the study's design. Those who supported the Aceh study led by Sommer convened several meetings including the National Academy of Sciences–sponsored Subcommittee on Vitamin A Deficiency Prevention and Control in 1986 (see National Academy of Sciences 1987). These meetings were effective stages for Sommer to recruit other scientists and policymakers in support of the charisma of vitamin A. Through these processes, the vitamin A pill as a magic bullet came to attain the status of scientific consensus.[11]

Another crucial element in the spread of vitamin A's charisma was the *political* appeal of Sommer's research, which emphasized the connection between vitamin A and *children's survival* rather than adult wellness. It is useful to note that until his study, vitamin A deficiency had been primarily considered an ophthalmological health issue, due to the deficiency's clinical manifestation in xerophthalmia or dry eye syndrome.[12] In contrast, vitamin A promoters in the 1980s linked it to "child survival." As Sommer told the *New York Times,* "When the main concern was night blindness, health ministers said, understandably, 'I feel terrible about that, but I can't put my resources into it when half our children are dying before the age of 5'…but now, ending the deficiency is starting to be viewed as a mainstream activity, not a peripheral one" (Eckholm 1985). Furthermore, because it explicitly benefitted children, who had a designated custodian in UNICEF with a mandate for the "survival, protection, and development of children" (United Nations 1992a, 140) vitamin A could add that powerful international institution to its stable of supporters. Indeed, UNICEF's head, James Grant, became especially known for his advocacy for vitamin A (Underwood 2004). He became famous for carrying vitamin A supplements in his pockets to use to tell stories about how these small pills could save children's lives. Sommer and Grant collaborated well together to cement political support for vitamin A, eventually securing US government funding for vitamin A pill distribution in developing nations under the category of "child survival."[13]

Charisma works magic, and vitamin A did not have a shortage of associations that suggested its magical power. Sommer and UNICEF's Grant emphasized the amazing potency of the small golden pill. Deploying uncharacteristically strong words for an established scientist, Sommer described vitamin A's impact as "absolutely unreal," and suggested that the improvement in child mortality was "in the order of 50 to 70 percent" (Rovner 1986). His zealous claims often irritated other experts. One researcher commented: "I wish he had not made such a high claim.…I don't think it's borne out in his study. A 10% claim would be more realistic. If his claims don't bear up in other studies, he could become the Linus Pauling of vitamin A" (Chris Kjolhede, quoted in Edmunds 1989, 14). Yet the seeming magic was part of the powerful image of vitamin A that circulated among experts and policymakers.

Emerging at the right moment, when the nutritional science community was searching for an exclusively nutritional contribution in international development, and blessed with powerful institutional and individual sponsors, vitamin A became charismatic in the 1980s. Its charisma was further strengthened as it developed links to life-or-death matters, to the most vulnerable group of society, children, and to the imagery of "absolutely unreal" potency for saving their lives. Vitamin A was tasked with the grand mission of saving children in the Third World, and the experts were prepared to provide a quick, easy, and cheap nutritional fix. Through the development of impressive institutional and personal sponsors that authorized and reified the message that vitamin A was a "magic bullet," the vitamin's scientific value was effectively translated into political and social values.

Women: In the Shadow of Children

As the historian of food Warren Belasco observes, the "starving children" of the developing world have been an icon of the world food problem (Belasco 2006). Any construction of a problem and its solution comes with an identification of "victims" of the problem and "beneficiaries" of the solution. The charismatic powers of both protein and vitamin A were critically linked to their victims/beneficiaries—children. This is most visible in the story of vitamin A, but protein was also often understood to be a children's problem, as kwashiorkor usually affected young children. Pictures of babies with pot bellies potently symbolized the centrality of protein to the health of children and the food problem.

What is often neglected in studies of food insecurity is how such attention to children also brings with it an incessant scrutiny of mothers. Such hypervisibility of mothers dawned on me when I observed a small neighborhood festival in Jakarta. It was a festival to promote health in a slum area. In their khaki uniforms, the officials from the health department marched into the canopy set up in a field in the neighborhood. Several men and women from the neighborhood lined up on the corner of the street to politely greet the officials. The officials were then seated in the first several rows of chairs in front of the makeshift stage where children sang songs for them. Then health workers ushered mothers into a line so they could put their children in the big sack of a hanging scale to weigh them. The mothers were in front of everyone where it would be revealed whether their child had "sufficient" growth. Perhaps because I had learned how some mothers in the district were marked in charts by health workers as "mothers with malnourished children," the weighing seemed like a mothering contest, with mothers judged on their feeding skills. To avoid child malnutrition, health workers emphasized, mothers had to be "aware" (*sadar*) of nutritional science. Mothers were seen as the key to the nutrition problem.

A growing body of work in feminist studies explores the relationship between medicine, health policies, and women. These studies have shown that there is a real possibility of *decreased* empowerment for women when their visibility is heightened in scientific and medical discourses focusing on their reproductive role. The increasingly pervasive mantra of child protection and, more recently, fetal protection has prompted medical experts to consider women solely in relation to children. In medical and policy discourses, the assumption of the mother-child dyad is frequently presented as a scientific necessity, yet feminist scholars have found that it is frequently accompanied by growing surveillance on maternal conduct and intrusion into women's bodies. The most striking cases involve the imposition of medical treatments (Ratcliff 2002) and criminal proceedings against pregnant women for causing fetal harm with alcohol (Gavaghan 2009) and drugs (Paltrow 1999). Treatments and punishments are imposed on women in the name of the child (Chase and Rogers 2001). Another example is the 2005 recommendation of the US Surgeon General that *all* women of child-bearing age abstain from alcohol as "potential mothers" (Gavaghan 2009).

Far from being a proportionate relation, the mother-child dyad frequently results in a mother's subordinate position in relation to her child/fetus that I call an *asymmetrical mother-child dyad*. This situation can become part of the story of charismatic nutrients when the mother's health is seen primarily as a means to her children's health. For instance, during the protein era, the role of women became more salient when experts shifted their focus from protein deficiency in school-aged children to that in preschool children. This meant that breast-feeding started to figure centrally in scientific debates, and breast-feeding practices came under increasing scrutiny by experts. Yet ironically, it did not mean that experts were concerned about the well-being of women. To a large degree, it was the breast milk that mothers produced that fascinated experts. For instance, experts were worried that women's growing employment outside the home might lead to their reluctance to breast-feed properly. An influential nutritionist who was active in promoting breast-feeding, D. B. Jelliffe, expressed his concern that "dedomestication of women" would decrease breast-feeding and increase formula feeding in developing nations (Ruxin 1996, 233). The celebration of women's breast milk was not a celebration of women's empowerment since women's reproductive role was prioritized over other roles. Women's complex decisions about choice and duration of breast-feeding was ignored.

Similarly, when experts worried about children's malnutrition it was rarely translated into advocacy for mothers. Often times, mothers' nutrition per se mattered little. As the Joint FAO/WHO Expert Committee on Nutrition flatly stated, "Malnutrition in mothers has been considered rather as a factor contributing to malnutrition in children *than as a particular problem in itself*" (quoted in Ruxin 1996, 90; my emphasis). This tendency to ignore women's health was perhaps exacerbated by findings that a mother's health status did not have a significant impact on the protein composition of the breast milk she produced (Belavady and Gopalan 1959; Ruxin 1996, 123).[14] As Beall (1997) notes about a more general trend in international health policy, policies are pursued "at the expense of women who are required to spend time, energy, and resources...often at expense to themselves" (79) and without much heed for nonmothers, such as elderly women.[15]

It is also important to note that the asymmetrical mother-child dyad highlights not only the mother's indispensable role for the child, but her inadequacy as a mother. The call to save children from malnutrition that accompanies charismatic nutrients often has resulted in implicit condemnation of women as ignorant, indifferent, and negligent in providing what is needed. Prominent nutritionist and breast-feeding advocate Donald McLaren passed judgment that "the main reason for the illness and deaths of children is not this scarcity. It is ignorance of infant care and infant feeding" (quoted in Ruxin 1996, 159). As the naturalized caretakers of the victim/beneficiary, mothers have been central to many experts' understandings of the essence of food insecurity.

The visibility accorded to women by the construction of the dyad clearly resonates with the history of food reform as discussed in the previous chapter, in which women have been caught in a commendation-condemnation bind. While food reformers have celebrated women's potential for improving food and nutrition, the applause often has been accompanied by the notion that women's inappropriate mothering, feeding, and nurturing were the root cause of the problem. And history also demonstrates that condemnation is especially reserved for women of lower socioeconomic status. Food reform movements in developed countries have had a tendency to single out for criticism mothers in immigrant, poor, and ethnic minority communities, rather than well-educated white mothers with economic means (see, e.g., Litt 2000). Women in developing countries also figure as "undesirable" mothers, although experts have had to simultaneously acknowledge their indispensable role in children's welfare.[16]

The profound irony of charismatic nutrients is that they tend to lead to casting the responsibility for malnutrition on mothers, but such realization does not inspire experts to collaborate with women to tackle the problem. Experts might have realized that when included in the conversation, women probably would undermine their "expert" recommendations: What if they were to say, "Please

give us decent work and housing before spending so much money on these cookies made from fish"? Indeed, the health-promoting festival mentioned earlier is
symbolic of the relationship between mothers and experts. Officials were there to
"give guidance" to mothers because, in their view, mothers might otherwise fail to
breast-feed or cook nutritious meals, thus jeopardizing the future of the nation.
They were not there to listen to mothers in order to collaborate on improving
children's health. Instead of seeing women as the agents of policy, experts tend to
prescribe nutritional fixes. By offering protein-rich engineered food, vitamin A
pills, and micronutrient-fortified food, experts have dodged the question of why
the women they condemn were unable to eat well during pregnancy, breast-feed
their babies, or cook nutritious meals. Women, overshadowed by the attractive
fixes, have been condemned as the agents of malnutrition but not trusted to be
the agents of improvement. Various magical fixes are delegated to solve food
insecurity, not women.

Selling Nutrition and Nutritional Fixes

We have seen how micronutrients in the 1990s were not the only instance of nutritionally driven interventions into the problem of food insecurity in developing
countries. Indeed, discussions of protein requirements and vitamin A bring on a
feeling of déjà vu that is hard to ignore. Despite apparent differences, the lack of
protein or the lack of vitamin A share characteristics with a micronutrient-based
diagnosis of the food problem. Privileging a particular substance as defining the
problem (charismatic nutrients) and providing solutions that are highly simpli-
fied (nutritional fixes) has been a constant theme in the history of global food
interventions.

Given the ephemeral nature of the reign of each charismatic nutrient and
nutritional fix, it is hard not to ask why they keep emerging. What do charismatic
nutrients do? Of course, they are supposed to fill the nutritional gaps and address
inadequacy in Third World food. But what kind of social work do they do? To
answer this question, one needs to understand that at the most fundamental
level a charismatic nutrient's critical function is to define the food problem as a
problem for the discipline of nutritional science to handle. Although the change
of diagnoses can be confusing, the discourse of charismatic nutrients, such as
those of protein and vitamin A, implicitly asserts a *nutritional* framing of the
world food situation. The institutional identity of the so-called Third World food
problem is quite ambiguous, more so than those addressed by immunization
(health) and illiteracy (education) campaigns, for example. In contrast, the Third
World food problem is not automatically strictly a "nutritional problem" or for

that matter an "agricultural problem," as food can be seen as belonging to over-lapping jurisdictions, including agriculture, population, and nutrition. It is in this context that charismatic nutrients help nutritional experts to mark the food problem as one that merits their expertise.

Charismatic nutrients' boundary-making function has been helpful for those in the field of international nutrition who have experienced their own insecurity as to their position in the scientific and development community. Importantly, the field's marginalization is related to the gendered history of nutritional science. A field traditionally dominated by women, nutritional science as an academic discipline has struggled with lack of respect, legiti-macy, and resources throughout its history (Apple 1997; Stage 1997; Levine 2008). Historian of nutritional science Rima Apple (1995) points out that nutritional science was long linked to the ideology of "scientific motherhood." This ideology prescribed that women need scientific knowledge to be success-ful mothers. It fuelled women's interest in nutritional science as well as society's desire to create an academic field to provide women with good homemaking skills, including the ability to prepare nutritious food. Considered one of very few "appropriate" academic fields for women, nutritional science came to be recognized as a "women's discipline."[17] However, its designation as belonging in the women's realm severely crippled it as a discipline. It suffered from lack of funding and was forced to concentrate on practical concerns and subjects readily available for study, rather than pursuing more prestigious "basic sci-ence" (Apple 1997). Many home economics departments operated as part of extension services and were expected to provide practical courses to girls so they could succeed in homemaking and child rearing. Marked as a "women's field," nutritional science "lacked the esteem accorded other departments that were composed of men and were considered more 'academic'" (30). As feminist scholar Sarah Stage summarizes, "Home economics . . . could never define itself outside of gender stereotypes" (1997, 12).

The gendered nature of nutritional science has been no less stark in devel-oping countries. Nutritional fieldworkers trained by colonial governments and international organizations also have been predominantly women (Ruxin 1996, 72; Calabro, Bright, and Bahl 2001). Although nutritional science in the West has gradually enhanced its cultural status by its link to chemistry and biol-ogy, nutritional science in developing countries rarely has been considered a prestigious scientific career (Ruxin 1996, esp. the excerpts from the interview with Scrimshaw at 67).

In addition, nutritional science has had trouble asserting itself in the exclusive circle of international development. Nutritional science was dwarfed by other disciplines in international organizations. For instance, at the end of the 1950s, international organizations had a very small number of nutritional experts: eighteen for FAO, one for UNICEF, and three (plus some more consultants) for WHO (Ruxin 1996, 111). Even in the 1980s, a Ford Foundation official, Lincoln Chen (1986, 71), offered a rather bleak assessment of the nutritional field in relation to international development:

> Nutrition does not command the excitement of research frontiers in the "new biology," nor does it compete in global significance with international economic relations. In many academic centers, nutritional interests have declined, owing in part to funding cutbacks....The nutrition community can no longer agree even on the magnitude of the global problem. Estimates of the world's malnourished range from 350 to 1,200 million. Controversy surrounds the food intake necessary to satisfy minimal requirements...there is also debate over the use of physical growth as a measure of malnutrition. With the knowledge base fundamentally so unstable, the nutrition community appears to be rudderless and to have little to offer in furthering understanding or problem-solving.

This quotation captures the perceived lack of legitimacy of the discipline of nutritional science in the realm of international development. In such a milieu, nutritional experts were compelled to create a tangible link between nutrition and development, and so the *nutritional* diagnosis of the food situation in the global South was valuable for asserting the relevance of the discipline to international development. To borrow Chen's words, charismatic nutrients helped the nutritional community to "compete in global significance."

For a feminized discipline struggling to gain respect within academia and in the field of international development, charismatic nutrients were strongly beneficial to its claims for legitimacy. As scholars of science and technology studies have pointed out, identification of an artifact specific to an academic discipline greatly enhances its stature and stability (Star and Griesemer 1999; Fujimura 1992).[18] For nutritional science, "nutrients" became the artifacts that drew the boundaries of the discipline and asserted its unique contribution and authority within the bounded space. Therefore, although the successive emergence (and disappearance) of charismatic nutrients that we have seen in this chapter might at a glance seem to indicate a disciplinary fracture, it actually worked to reinforce the discipline's claim of the nutritional character of the food problem. Different nutritional scientists might have been committed to different nutrients, methodologies, and solutions, but the nutritional community as a whole shared a stake in insisting on a nutritional representation of the Third World food problem.

Situated in the broader politics of academic disciplines and international development, the institutional and cultural appeal of nutritional fixes becomes clearer. To sell nutrition, nutritional fixes such as super protein cookies and vitamin A pills were critical. The palpability of the solution symbolized in these fixes was important when the nutritional community had to market food-and-nutrition related projects to governments and international organizations. Food policy experts had always competed with those advocating other development projects that might be more obviously rewarding to the recipient governments. For instance, FAO's nutritionist, Jean Ritchie, complained that "in the minds of the Public Health Departments and Governments in general the UNICEF's supplies of D.D.T, dried milk etc. are associated with WHO, who get credit for bearing gifts with them. Until we have something to offer in the way of laboratory equipment or other such supplies associated with TC [technical assistance] personnel, the competition will be tough" (quoted in Ruxin 1996, 101). It was this need to sell nutrition in tough competition with other sectors that nutritional fixes effectively assisted. Recall, for instance, the case of vitamin A that was promoted as a "dirt cheap" golden bullet (Brown 1994). Like Sommer, who always emphasized that vitamin A supplements were the "cheapest, most practical means of increasing childhood survival"(quoted in *Newsweek* in 1985, cited in Edmunds 2000, 20), the nutritional community needed a cheap, practical magic bullet to sell nutrition to developing countries and development organizations, and these attributes were at the core of the attractiveness of charismatic nutritional fixes.

Charismatic nutrients conjure up scientific facts, ethical judgments, and the promise of solutions. Their emotive power is undeniable: they tell stories about babies with swollen bellies who do not have eggs or milk, about golden pills that save children and cost only a few cents, and about developing countries' "lost generations"—lost due to the invisible nature of micronutrient deficiencies. Their validity is solidified through scientifically determined nutritional "needs" that concretize the notion of "gaps" in nutrients. Charisma, however, often has been fragile, particularly when the therapeutic efficacy of technical fixes—which were expected to offer magic cures—failed. Yet the next charismatic nutrient is always waiting, as having such icons is crucial for the legitimacy, prestige, and vitality of the nutritional community, which has been particularly handicapped in its competition with other disciplines in international development partly due to its historical feminization. But what is left in the dark when charismatic nutrients fills the limelight? While charismatic nutrients and their attendant fixes produce useful justifications for nutritional experts to claim, protect, and advance their sector and career, they also lead to a critical absence of attention to non-nutritional issues. By defining the problem as a "gap" in certain charismatic nutrients, other important gaps—say, gaps in men and women's social power, land ownership, income, education, and unionization—are ignored. They silence other possible ways of articulating problems by closing the frame of understanding tightly around an increasingly small space. Stealing the stage with their charisma, select nutrients become the only face of the Third World food problem.

3

SOLVING HIDDEN HUNGER WITH FORTIFIED FOOD

> **The World Bank is calling for the inclusion of nutrition schemes in every appropriate Bank project in order to combat deficiencies in Vitamin A, iodine and iron. Without these so-called micronutrients, development is hindered in many countries by the need to care for the more than 1m cases of blindness, mental retardation, learning disabilities and low work capacity, says a Bank report. The recommendation to devote more resources to these nutrients is contained in "Enriching Lives" published by the World Bank's human resources division, now the fastest growing sector in the giant multilateral lending institution.**
>
> —Nancy Dunne, *Financial Times,* December 17, 1994

Nutritional fixes such as protein-enriched food and vitamin A pills have had their day. The contemporary nutritional fix is *fortified food,* which became popular in the 1990s. The popularity of fortification is intriguing because there are many ways to conceive of solutions to micronutrient deficiencies. In the lexicon of nutritional science, "micronutrient strategies" refer to a set of public health interventions to combat micronutrient deficiencies. Typical micronutrient strategies suggested by international experts to Third World governments include supplementation and dietary diversification/education in addition to fortification (Underwood 1998; Maberly, Trowbridge, and Sullivan 1994).[1] The question is why fortification among these possible options? The answer lies not *within* fortification but *outside* and *around* it. Although the technical and biological merits of fortification tend to be seen as the reasons for its popularity, its political functions are no less important.

In this chapter, I address two important factors behind the fortification boom. The first is fortification's ability to simplify issues to technical matters, and consequently to avoid direct engagement with women. Women are the focus of micronutrient interventions because they are more likely to have micronutrient deficiencies and they influence children's micronutrient status during pregnancy and through nursing and feeding them. But convincing women to follow expert instruction and changing their behavior in terms

of diet and feeding practices presents a great challenge. In contrast, fortification does not involve convincing women to change their dietary and feeding
behaviors. Nutrients can be added to products that people usually consume,
and fortification can be designed exclusively by experts from the food industry, scientists, and policymakers. The second factor is fortification's fit with
neoliberalism. While fortification can theoretically be carried out by governments, micronutrients are typically added to existing products such as wheat
flour, formula milk, and margarine in the private sector. In the case of Indonesia, for instance, wheat flour fortification is mandated by the government, but
the actual process of adding vitamin premix to wheat flour and marketing and
distribution is carried out by the private sector. Voluntary fortification of baby
formula is similarly done by formula companies. In this chapter I describe how
multilateral lending institutions such as the World Bank and the Asian Development Bank (ADB) have played a critical role as the sponsors of fortification
projects.

Arturo Escobar has discussed the process of the "economization of life" (1996,
331) in which international development accelerates the power of rationality
and economic calculation espoused by development economists. Following his
concept, I use the term "economization of nutrition" to show that neoliberalism
has brought forth a particular framework of analysis and diagnosis that casts
an economic gaze on food, health, and nutrition. The economization of nutrition refers to the processes through which capitalist economics is increasingly
introduced into world nutrition problems. Explicit reference to theories and
the use of calculative tools from conventional economics to gauge the efficacy
of programs is an important epistemological requirement imposed by these
multilateral banks on the global food and nutrition policy community.[2] One of
the important consequences of this has been the construction of fortification
as superior to other micronutrient strategies. By changing the interpretation of
"advantages" and "benefits" of different micronutrient strategies, the economization of nutrition has critically shaped actual interventions into the food and body
of the Third World.

These two issues—the absence of women and the prominence of the market under neoliberalism—are tightly interrelated. Experts' suspicion of women's
ability to eat and feed properly has helped to legitimize market-based, corporate-
centered strategies. While the market emerged as the provider of the solution
to micronutrient deficiencies, women became merely victims of micronutrient
deficiencies, and hence passive recipients of fortified food, rather than being
considered as part of the solution. Such a gendered assessment of strategies to
address problems of food security is linked to particular moments in international development.

Discovering "Hidden Hunger"

In the 1990s, micronutrients emerged as the new charismatic nutrients, and "micronutrient deficiency" became a central focus of the international nutrition community. The charisma of micronutrients was solidified and amplified by multiple international conferences that advanced micronutrient deficiency as the major problem of Third World food. These conferences include the 1990 World Summit for Children, the 1991 "Ending Hidden Hunger" conference (convened by the same group), the 1992 International Conference on Nutrition (ICN), and the 1996 World Food Summit (Underwood and Smitasiri 1999).[3] Pledging "to make all efforts to eliminate before the end of this decade...iodine and vitamin A deficiencies...and other important micronutrient deficiencies, including iron" (FAO and WHO 1992), these conferences set specific micronutrient goals to be addressed by global society. For instance, the 1990 World Summit for Children agreed to eliminate iodine deficiency disorders (IDD) and vitamin A deficiency (VAD) by the end of the century and to reduce iron deficiency anemia by one-third from 1990 levels. The charisma of micronutrients was further fortified with their incorporation in the Millennium Development Goal (MDG), which was agreed to by all the member states of the United Nations in 2000. The importance of micronutrients was similarly emphasized by the "World Fit for Children" declaration, endorsed by the 27th Special Session of the UN General Assembly in May 2002 (Rogers 2003). The World Food Summit in 2002 also targeted ending micronutrient deficiency as a goal, reflecting the growing international interest in micronutrients:

> We emphasize the need for nutritionally adequate and safe food and highlight the need for attention to nutritional issues as an integral part of addressing food security....We recognize the importance of interventions to tackle micro-nutrient deficiencies which are cost-effective and locally acceptable. (FAO 2002a, 85)

The new charismatic nutrients started to attract a number of micronutrient-focused projects by international organizations. For instance, USAID started the Opportunities for Micronutrient Interventions (OMNI) project in 1993 to help countries meet the goals set by the World Summit for Children and ICN. Similarly, participants at the "Ending Hidden Hunger" conference established another organization called the Micronutrient Initiative (MI) to promote micronutrient strategies in developing countries. Another organization, Global Alliance for Improved Nutrition (GAIN), was established by the Bill and Melinda Gates Foundation, USAID, CIDA, and the World Bank to promote micronutrient awareness in developing countries. There have also been various coordinating programs for micronutrient

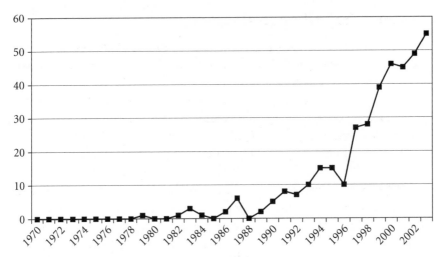

FIGURE 3.1. The number of publications with key word "micronutrient malnutrition."
Source: PubMed, National Library of Health.

strategies, such as the Program Against Micronutrient Malnutrition (PAMM) at Emory University, which pays particular attention to IDD and fortification.[4]

The ascendancy of micronutrients into stardom in international development was accompanied by their rising popularity in academia. A search in the database of medical and nutrition journals at the National Library of Medicine (PubMed)[5] by the keyword "micronutrient malnutrition" indicates that the bulk of academic publications on micronutrients have been published since 1990 (fig. 3.1). The impressive list of prestigious global conferences and the rising number of academic publications attest to the rise of micronutrients as the new charismatic nutrients of the 1990s and into the 2000s.

Fortification on the Rise

When I did extended field research in 2004–5 in Indonesia, fortification was on many organizations' agendas. I interviewed international organizations, donor agencies, NGOs, nutrition and food policy experts, and government officials, and everyone seemed to be talking about their new "smarter" food products to be distributed in the country (see table 3.1). "Lost generation" and "hidden hunger" were catch phrases used by many whom I interviewed to communicate the dire yet little known consequences of micronutrient deficiencies and to shore up political support for fortification projects.

TABLE 3.1 Examples of international organizations' nutritionalized projects in Indonesia as of 2004

ORGANIZATION	PRODUCT TYPE	STARTING YEAR	CHARACTERISTICS
UNICEF/WFP	porridge	n/a	economic crisis emergency
Mercy Corps	porridge	n/a	economic crisis emergency
UNICEF/WFP	addition to porridge for children under two	2000	economic crisis emergency
HKI	sprinkles	n/a	experiments
IRD (USDA title II)	instant noodles	n/a	business development
Land O' Lakes	milk	1998	school feeding
WFP	cookies, instant noodles	2004	urban nutrition, postemergency
Mercy Corps (USDA)	soy milk	n/a	school feeding
government of Indonesia	complementary food	2001	nutrition program

The most visible example of a fortification program in Indonesia was food aid in conjunction with the Asian financial crisis that started in 1997. International donors from USAID to UNICEF distributed various fortified products.[6] For instance, the World Food Programme distributed a fortified complementary food for infants called Vitadele as emergency relief.[7] The program was started in 1999 and expanded in 2000 to cover 375,000 young children. Unfortunately, Vitadele's evaluation showed that mothers did not like the product because it had to be cooked with other food and took too much time and effort. The WFP then introduced a new fortified complementary food called Delvita. Delvita is a sachet of "sprinkles" containing microencapsulated iron and other micronutrients (Soekirman et al. 2005). The WFP distributed this product in urban Java, including Jakarta, Bandung, Semarang, and Surabaya, between 2000 and 2003. In 2004, although the most acute phase of the economic crisis was over, the WFP began another fortified food project to distribute fortified cookies made by Danone and instant noodles made by Indofood.[8]

It was not only international governmental organizations that were riding the fortification wave. Fortified food also was increasingly popular with international NGOs that worked in Indonesia. In 1998, Land O'Lakes started a school lunch program in Indonesia using fortified milk and instant noodle snacks. Another relief NGO, the Mercy Corps, distributed fortified Vitadele and soymilk with USDA funding.[9] A US-based nonprofit organization, International Relief and Development, started making fortified instant noodles using wheat flour donated under USDA's Title II program called Food for Progress. They also started to make noodles made from soy flour, and when I interviewed him, the program officer was excited about another new product of fortified rice noodles and sweet soy sauce.[10]

The popularity of fortification is also evident in that an NGO that used to emphasize supplements is now experimenting with fortification. Helen Keller International (HKI) is an international NGO that has been particularly active in working with the Indonesian government to distribute vitamin A capsules since the 1970s.[11] But in recent years, HKI has started to emphasize fortification as well and has launched a fortification product called Vitalita. Vitalita is a sachet of multivitamin sprinkles that can be added to homemade baby food. It is manufactured by a multinational food producer, Heinz.[12]

In addition, the Indonesian government itself started a fortified food project. Since 2001, it has allocated the bulk of its nutrition budget to food assistance with fortified food. Fortified with micronutrients, the distributed baby food is made by a food conglomerate, Indofood. Initial funding came from the ADB, but after it ended, the government continued the distribution with its own money. The emphasis on this project was enormous; the budget for this fortified baby food program amounted to 65–70 percent of the total national budget for nutrition programs in 2002–4 (Soekirman et al. 2005). The government also passed Indonesia's first mandatory fortification law.

The Indonesian situation reflects how fortification became popular in the international nutrition community and involved scientists, bureaucrats, NGOs, and food companies. Following the global declarations and agreements on micronutrients, the international development community started a number of micronutrient-focused projects, many of which were fortification projects. The Global Alliance for Improved Nutrition and the Micronutrient Initiative are good examples. GAIN was funded by the World Bank and other organizations with a specifically fortification-related mission: to build "momentum to end vitamin and mineral deficiencies through the fortification of staple foods and condiments" (GAIN 2005). Even non-nutrition NGOs have been lured into the fortification enterprise. In 2006 Grameen Bank and multinational food manufacturer Danone entered a joint fortification venture, Grameen Danone Foods, to produce fortified dairy products. Grameen Bank is, of course, the well-known pioneer of microfinance whose founder, Muhammad Yunus, received the Nobel Peace Prize in 2006. The joint venture's yoghurt, Shakti Doi, is fortified with micronutrients such as iron, zinc, and calcium.[13]

Fortification has also been incorporated as a national program in many countries (see table 3.2). In accordance with the "global consensus" outlined above, a number of developing countries have started national mandatory fortification programs with various vehicles since the 1990s (Darnton-Hill and Nalubola 2002). Various developing countries now mandate fortification of different food items, such as iodization of salt, iron fortification of flour, and vitamin A fortification of sugar, oil, and margarine.[14]

TABLE 3.2 National fortification projects in developing countries

COUNTRY	ITEM	THIAMINE	RIBOFLAVIN	NIACIN	FOLIC ACID	FE	VIT A	VIT D	CA	ZN
Bolivia	wheat flour	X	X	X	X	X				
Brazil	dried skimmed milk						X	X		
Chile	wheat flour	X	X	X	X	X				
	pasta	X	X	X		X				
	margarine						X	X		
Columbia	wheat flour	X	X	X	X	X				
	margarine						X	X		
Costa Rica	wheat flour	X	X	X	X	X				
	sugar						X			
Dominican Republic	wheat flour	X	X	X	X	X				
Ecuador	wheat flour	X	X	X	X	X				
	margarine						X	X		
El Salvador	wheat flour	X	X	X	X	X				
	margarine						X			
	sugar						X			
Guatemala	wheat flour	X	X	X	X	X				X
	pasta	X	X	X		X				
	skimmed milk						X	X		
	margarine						X			
	sugar						X			
Honduras	wheat flour	X	X	X	X	X				
	milk						X	X		
	margarine						X	X		
	sugar						X			
Mexico	milk						X	X		
	margarine						X	X		
Nicaragua	wheat flour	X	X	X	X	X				
	sugar						X			
Panama	wheat flour	X	X	X	X	X				
	sugar						X			
Paraguay	wheat flour	X	X	X	X	X				
Peru	wheat flour					X				
	margarine						X	X		
Venezuela	wheat flour	X	X	X		X				
	precooked maize meal	X	X	X		X	X			
	dried milk powder						X	X		

(Continued)

TABLE 3.2   *(Continued)*

COUNTRY	ITEM	THIAMINE	RIBOFLAVIN	NIACIN	FOLIC ACID	FE	VIT A	VIT D	CA	ZN
Nigeria	flour	X	X	X		X			X	
South Africa	maize meal		X	X						
	margarine							X	X	
Zambia	sugar						X			

Source: Darnton-Hill and Nalubola 2002.

Sponsors for Fortification

Who were the critical sponsors for fortification as the new nutritional fix? As we have seen, charismatic nutrients and attendant nutritional fixes often have powerful institutional connections and sponsors who effectively propagate the efficacy and charisma of a particular nutrient and the moral imperative for adopting a particular nutritional fix. In the case of fortification, the role of the World Bank has been crucial.

Needless to say the rise of fortification is attractive to the food industry. From the perspective of the food industry, the international embrace of fortified food means that products can be marketed as "healthy" and "necessary." This explains, for instance, why the International Life Sciences Institute has been particularly active in promoting fortification in developing countries. Although ILSI publishes a journal called *Nutrition Review* and looks like an independent academic research institute, it is actually an organization funded by the food industry with major transnational companies such as Nestlé and Kraft as its members. ILSI has hosted various workshops on fortification, often in collaboration with international organizations in developing countries. The food industry not only welcomes the spread of fortification advocacy, it also wants to shape the fortification policies in developing countries so that fortification standards are harmonized to ease the penetration of Third World markets. As one of the industry people I interviewed put it, "Corporations want the global recipe."

The food industry has not been the only engine behind fortification. The epigraph at the beginning of this chapter is revealing if we notice not only *what* was being called for (the micronutrient strategies), but also *who* called for it (the World Bank). Among international organizations, the World Bank has had a crucial role in seeing fortification as the "solution" for food problems. As is perhaps already is obvious from the above description of international micronutrient projects, many of which had the World Bank as a partner, the World Bank has been particularly central to the international fortification network. By the 1990s, the World Bank explicitly expressed its commitment to micronutrients in its iconic publication, *Enriching Lives,* which was solely devoted to the analysis of micronutrient deficiencies in developing countries and argued that "the control of vitamin and mineral deficiencies is one of the most extraordinary development-related scientific advances of recent years." The text promoted micronutrient strategies, stating that "probably no other technology available today offers as large an opportunity to improve lives and accelerate development at such low cost and in such a short time" (1994, 1). It even sought to include a micronutrient component in any World Bank project implemented in countries with such problems (Dunne 1994). The World Bank has become a formidable powerhouse in pushing the fortification agenda in international development. The international institutional networking for fortification has depended on the World Bank's resources. For instance, the Bank was the key founder of the Micronutrient Initiative and GAIN. In addition, the Bank started the Business Alliance for Food Fortification in 2005, which is a partnership with the private sector to promote fortification. BAFF partners with the major players in the global food industry including Nestlé, Heinz, Ajinomoto, Dannon, and Unilever, and is chaired by Coca-Cola (GAIN 2005). Insisting that food fortification is "one of the most promising interventions for improving the nutritional status of the world's poorest and should be the first area of focus" in nutrition policy (BAFF 2005), BAFF campaigns for private-public partnerships for fortification in developing countries.

Another powerful multilateral lending institution, the Asian Development Bank, has also sponsored fortification initiatives. Hosting many conferences and workshops, it has been critical in the promotion of fortification in Asia. For instance, the ADB convened a regional fortification conference in Manila in 2000, which was cosponsored by the International Life Sciences Institute and the Micronutrient Initiative (ADB 2000b). After this forum, ADB, ILSI, and the Danish International Development Agency started technical assistance programs in six countries that examined ways to encourage food fortification by the private food industry (ADB 2000c). In addition, the ADB has hosted various fortification workshops and meetings such as its Workshop on Flour Fortification and Workshop on Cooking Oil Fortification in 2001 in New Delhi, its Workshop on Complementary Foods Technology and Workshop on Infant Feeding Practices in 2001 in Singapore, its Regional Dialogue on Food Fortification, Trade, and Surveillance in 2001 in Thailand, and the Investor's Roundtable in 2001 in Shanghai (Hunt 2001a).

Such a leadership role by multilateral lending institutions in the area of nutrition begs the question of their motivations. Why did they particularly find fortification a worthy project for their support and advocacy? This question has to

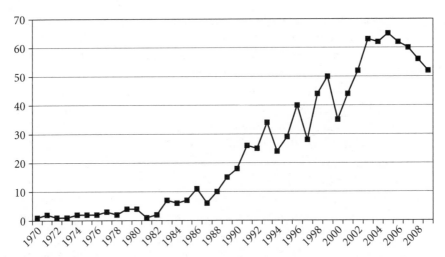

FIGURE 3.2. The number of newly approved World Bank projects with Health, Nutrition, and Population code.

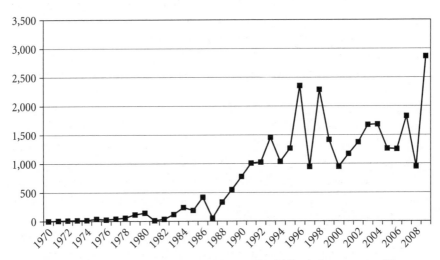

FIGURE 3.3. HNP sector commitments by the World Bank (in current millions of dollars).

be considered in a broader context of shifting involvement in the health and nutrition sector by these banks. In traditional development economics, health and nutrition did not count as a development program; they were thought of as a cost rather than an investment for development. Hence the multilateral lending institutions used to focus on large-scale infrastructure projects such as the dams, highways, and ports that were viewed as essential for "economic development" in the traditional sense as measured in terms of GDP growth. Health and nutrition were not their priorities (Fair 2008). However, the importance of social projects was gradually realized. Theories of development started to emphasize the importance of "human capital" and "human development" as early as the 1960s. There was also an increasing amount of research that showed the economic consequences of malnutrition. Studies started to document productivity impacts of malnutrition (Basta et al. 1979; Karyadi 1973b; Alderman, Hoddinott, and Kinsey 2006). It became increasingly clear that an economic rationale could be made for nutritional lending.

Scholars have suggested that another factor might be structural adjustment policies (SAP) (see, e.g., Baru and Jesani 2000). When many developing countries faced massive debt problems in the 1980s, the multilateral lending institutions imposed strict conditions on assistance, such as currency devaluation and trade liberalization. In addition, they often called for public sector contraction, including privatization of state-owned enterprises and cuts in government jobs and social programs. Social activists saw this as a serious attack on social projects and criticized the World Bank for what they saw as the neglect of social development (Rich 1994). And the critique was not limited to radical social movement activists. The prestigious medical journal, *The Lancet,* for instance, had an unusual editorial in 1990 that denounced the contradictions of the SAP policies that forced cuts in government expenditures while recommending improvements in health services (Lancet 1990). Even international organizations such as UNICEF joined the critics, asking for "adjustment with a human face" (Cornia, Jolly, and Stewart 1987). UNICEF estimated that SAPs were associated with the deaths of five hundred thousand young children in a twelve-month period (UNICEF 1989, 16–17).[15] These negative health impacts of SAPs helped to accelerate a process of increasing funding for health and nutrition by the World Bank (Baru and Jesani 2000; Levinson 1993).

The multilateral lending institutions' engagement with issues of health and nutrition was at one time relatively insignificant. It's true that the World Bank had produced several publications on the subject of nutrition (Karyadi 1973b; Reutlinger and Selowsky 1975), and there was some nutrition-related lending. For instance, the Bank provided funding for UN nutrition activities, such as the Protein Advisory Group that was established in 1955 and the Administrative Committee on Coordination, Sub-Committee on Nutrition that was established in 1977. Alan Berg, who promoted multisectoral nutrition planning that was discussed in chapter 2, was a nutrition consultant for the Bank. Nonetheless, it was only in the 1990s that multilateral lending institutions became active sponsors of health and nutrition projects. The number of projects and the lending amount for the Bank's Health, Nutrition and Population (HNP) sector has increased steadily since the 1990s (see figs. 3.2 and 3.3). HNP was only 1.6 percent of total World Bank lending in 1984, but it is now close to 11 percent.[16] In the 1990s, the World Bank even came to be considered "a heavyweight in international health" (Lee et al. 1996).

As the World Bank's involvement with health grew, it increased its commitment to nutrition as well. Nutrition was one of several social sectors that the Bank had previously avoided. Until the early 1990s, its investment in nutrition programs remained quite low, averaging only $16 million per year (World Bank n.d.). Yet the commitment increased, and the Global Hunger Conference in 1994 and the Bank's *Strategy for Reducing Poverty and Hunger* (Binswanger and Landell-Mills 1995) even featured malnutrition as one of the Bank's critical mandates. The Bank's nutrition-related lending increased significantly to $140 million per year by 1995, and about half of its Social Funds projects included nutrition activities (World Bank n.d.).

However, there still remains a question: Of the myriad health and nutrition-related projects in developing countries, why is there particularly strong advocacy for fortification from the World Bank? Let us tackle this question by going back to the concept of economization of nutrition.

Nutrition as Investment and Malnutrition as Economic Loss

When I conducted interviews in Indonesia about fortification policy, it was hard to overlook a peculiar business-y vibe that was shared by many I talked with. Many were fluent in mixing development concepts with economic jargon to the extent that I thought they might have obtained an MBA before coming to the field of international development. Describing the fortification programs of his NGO as having "market multiplier effects, down the whole marketing chain, as well as generating income" and helping local factories increase operating capacity as well as having created "thousands of sustainable jobs," an American staff member was upbeat about these ostensible "nutrition" programs. Similarly, one staff member of an international organizations said of micronutrient programs, "We can easily calculate how much money we saved. It's in dollars....I tell them, for every one dollar you spend on iron deficiency you get a hundred dollars back. Much more." This might not have been surprising if the man were a development economist. But he was a physician who also had degrees in health and nutritional science. His use of good "return on investment" and "money saved" in describing micronutrient strategies indicated something was going on in the discourse around malnutrition in developing countries.

As one might expect, the World Bank's espousal of micronutrition is significant for financial reasons, because these multilateral lending institutions' financial power is immense (Kickbusch 2000; Lee et al. 1996). As Lee et al. (1996) note, the Bank's unrivalled financial resources, which are used to provide low-interest loans and credits, makes the World Bank tremendously important for developing country governments, international organizations, NGOs, and other development agencies. The influence of the World Bank is not limited to its financial muscle, however. The addition of nutrition to the Bank's portfolio has been important not only in terms of the political economy of nutrition but also in terms of social understandings of it. As Goldman (2001) points out, the World Bank has a significant *epistemological* power to impose certain policy assumptions and frameworks onto other organizations and governments. This makes it imperative to examine how the World Bank approaches nutrition issues. How does it represent the food problem and its proposed solutions in the Third World? What are the implicit links that the World Bank draws among health, economy, nutrition, and food? How does it describe the reality, and how does it prescribe solutions?

From interviews that I conducted and my analysis of the Bank's publications, it became clear that the Bank's sponsorship of micronutrients and fortification is fundamentally linked with its economized view of nutrition, and rationalized in terms of economic efficiency, relative economic cost, and economic loss and gain. The most obvious rendition of the economized view of nutrition is the concept of "disability adjusted life years" (DALYs) (Murray and Lopez 1999). DALYs are used to calculate the monetary cost of ill-health and monetary benefits of health interventions by assigning different values to life lost at different ages. The value for each year of life rises from zero to peak at age twenty-five and then declines gradually. DALYs provides a concrete numeric representation to the economics of micronutrients and hail its economic significance. For instance, the Bank asserted in its 1994 publication on micronutrient deficiencies, *Enriching Lives,* that "most micronutrient programs cost less than $50 per disability-adjusted life-year (DALY) gained. Deficiencies of just vitamin A, iodine, and iron—the focus of this book—could waste as much as 5% of gross domestic product, but addressing them comprehensively and sustainably would cost less than 0.3% of gross domestic product (GDP)" (2). Indeed, *Enriching Lives* is full of such an economized view of nutrition. Repeated throughout is the description of micronutrient deficiencies and strategies from the vantage point of economic calculation:

> No other technology offers as large an opportunity to improve lives…
> at such low cost and in such a short time. (cover)

The message is clear: the problem is huge, solutions are "on the shelf", and few countries can afford not to address micronutrient malnutrition. (cover)

The control of vitamin and mineral deficiencies is one of the most extraordinary development-related scientific advances of recent years. Probably no other technology available today offers as large an opportunity to improve lives and accelerate development at such low cost and in such a short time. (1)

Fortunately, all of these options are inexpensive and cost-effective. (2)

Micronutrient interventions are among the most cost-effective investments in the health sectors. (5)

The economic and social payoffs from micronutrient programs reach as high as 84 times the program costs. Few other development programs offer such high social and economic payoffs. (5)

Describing people as "consumers" and micronutrient strategies as "on the shelf," "cost-effective" and with high "payoffs," the Bank's interpretation of micronutrient deficiencies and policy options is resolutely grounded in the economized view.

While the idea that better nutrition leads to better productivity might not seem problematic, the underlying logic has profoundly disturbing assumptions. For instance, under the calculations based on DALYs, disabled or chronically ill people's lives are considered less valuable than those of normal people. Programs that do not result in cures or prevention can be viewed as too expensive. In addition, because of the way the DALYs are calculated, the very young, the elderly, and disabled people have little economic value. The calculation might conclude that very little of importance would be gained by addressing these people's needs (Murray and Lopez 1999).[17]

That the Bank's understandings of nutrition are based on economic calculations might not be surprising, but such understandings are increasingly powerful beyond the World Bank, as my interviews indicated. For instance, the Asian Development Bank's support for micronutrient projects takes place in a similar framework. The ADB maintains that micronutrient malnutrition is important because it "will cost the economy at least 3% of GDP annually," and food fortification must be promoted because it is the "most assured and least costly strategy" (ADB 2000b, 9). Even traditional health sectors such as WHO and UNICEF are not immune from such discourse. For example, UNICEF started a damage assessment report on malnutrition, which was a new effort to provide evidence that malnutrition costs money to a country's economy. Explicitly trying to put a price tag on malnutrition, the damage assessment report mirrors the DALYs and replicates its economic logic. As Goldman (2001) has astutely observed, the Bank's epistemological influence is indeed far-reaching.

The increasingly pervasive economized view of nutrition critically changes how solutions to the Third World food problem are evaluated. More specifically, it has influenced the hierarchy of different micronutrient strategies, putting fortification at the top. Fortification's ascendancy is illuminating when one considers that supplements used to be the most utilized option among the micronutrient strategies. Strongly advocated by major international health organizations, such as UNICEF and WHO, and NGOs, such as Helen Keller International, the supplementation approach was the mainstream choice to tackle micronutrient deficiencies in developing countries. For instance, the WHO issued a recommendation in 1987 to distribute vitamin A supplements in conjunction with a national immunization day, and many countries followed this plan (WHO 2003). For other micronutrient deficiencies, such as iron deficiency anemia and iodine deficiency disorder, pharmaceutical solutions were the standard practice, rather than fortification. Governments distributed iron tablets to pregnant mothers and iron syrup and iodine syrup to children. In the economized language of nutrition and health that is increasingly prevalent, however, supplementation is rendered problematic because of its heavy state involvement. The procurement and distribution of supplements requires too many government resources or those of international organizations, and the execution is dependent on their capacity and commitment. In contrast, the argument goes, fortification is much more "efficient" because it needs less government involvement.

Fortification is also an ideal way to involve the private sector in the currently celebrated notion of "public-private partnership." Public-private partnership promotes the collaboration between the government and the private sector for social projects (Maberly 2002). While it draws on historical examples of social reform projects by charity organizations, as Miraftab (2004) points out, the concept has gained strength under neoliberal ideology. Public-private partnership is now seen as a way to reduce the role of government and government expenditures on public services, replacing the state with private firms, which are deemed more efficient service deliverers. Based on the belief that the market is better equipped to offer solutions to social problems, the Business Alliance for Food Fortification is emblematic of the Bank's commitment to public-private partnership as the basis for food reform in developing countries. This statement by BAFF underscores such ideology of private sector partnership for public policy purposes that lurks behind the Bank's sponsorship of fortification in general:

> The role of the private sector in creating market-viable and sustainable food fortification is integral due to its strengths in products, technology and marketing. The poor in developing countries constitute the largest population in need of vitamins and minerals. If they are to be reached, the private sector's strengths must be tapped into and expanded and the challenges it faces must be voiced. (GAIN and BAFF 2005)

Celebrated as a tool to tap into the private sector's know-how and technology, fortification comfortably satisfies the parameters set by the economized view of nutrition. Making other options seem antiquated or just inarticulatable in the age of the mandatory neoliberalization, economization of nutrition critically informs the construction of fortification's "advantages" and "superiority" as an intervention strategy for hidden hunger.

Shaped by neoliberal ideology, the economization of nutrition has had tremendous influence on what is to be done about "the food problem." Fortification's ascendance has not been simply inevitable due to its "scientific" superiority as a nutrition solution, but instead it is intimately linked with the increasingly economistic framing of nutrition and health in international development, promoted by multilateral lending institutions whose influence in the health sector has grown tremendously. Normalized by the new giant in the international health field, the World Bank, and increasingly pervasive beyond it, the economized view of nutrition was critical to making fortification look ideal as a way to address hidden hunger.

From Women to Market

Scholars of food studies need to ponder two further implications of the rise of fortification. First is the understanding of the market as the solution, rather than the problem. Noticeable within the prevailing discourse of fortification is the belief that governments and international organizations can reap monetary savings from properly using market mechanisms and that private corporations can provide the most efficient solution to the problem of malnutrition. This particular problematization was made possible by nutritionism. With its exclusive focus on nutrients, nutritionism makes the market the ideal mechanism to channel nutrients to the mouths of consumers. Such a characterization of the free market and private industry fails to acknowledge that neoliberal economic restructuring and increasing global "free trade" regimes has decreased, rather than increased, people's stable access to food. In addition, if nutritionally necessary food is to be provided in the form of fortified food sold via commercial markets, what happens when they become unaffordable due to market fluctuations? The food crisis of 2007–8, when food price skyrocketed and food riots erupted in developing countries, is instructive. Among the hardest hit products were oil (used for cooking oil and margarine) and wheat flour, which were popular items in fortification programs. In Indonesia, for instance, within two months in early 2008, wheat flour prices increased by 15 percent (Meylinah 2008). Dependence on the market, and particularly on imported food for essential nutrition, is risky when one considers the volatile conditions of the market.

In addition, we have to note that the market's principal logic is profit and return on investment. Note what happened to the Indonesian fortification program during the 2007–8 food crisis. Citing the high price of wheat, the Indonesian milling industry lobbied the government to suspend the fortification requirement in order to allow for importing of wheat flour. As a result, the Ministry of Trade and Industry temporarily lifted the national fortification requirements on wheat flour in January 2008 (Meylinah 2008). It was precisely in this kind of crisis situation that fortification programs might be most helpful for the poor. However, when markets fluctuate violently and uncertainty about profitability gets heightened, corporations lose resources and motivation for fortification. Since by definition they must maximize profits for shareholders those goals come before the needs of the poor, malnourished citizens of developing countries.[18]

Dependence on fortification by the private sector leads to vulnerability to the volatility of global markets and corporate calculations of profitability. This casts significant doubt on a stabile market-based solution. The forces that promote fortification, however, tend to see the market solely as a solution, rather than as a problem. They continue to see the market as able to deliver missing micronutrients to the mouths of the poor in an efficient manner.[19]

The second implication, related to the first, is that as experts celebrate their new partnership with private industry, women have faded away as potential partners for solving the food problem. One important aspect of fortification's constructed advantage vis-à-vis other strategies is related to a particular view of women in the experts' discourses. Nonfortification strategies to combat micronutrient deficiencies, notably supplements and nutritional education, have been sidelined, not only because they do not fit the economizing vocabularies that have become increasingly powerful in the global domain of food, but also because of the view of experts that they have a severe "compliance" problem. For instance, supplements need to be taken by people once they are delivered. People need to be convinced of the need and efficacy of supplements. Compliance is particularly important in the case of iron deficiency anemia programs, since iron cannot be stored in the body and has to be taken regularly. Iron pills also can have numerous side effects, including stomach discomfort, which makes many women not take the pills as prescribed. Experts lament that women may either not take the pills at all or take them sporadically. The World Bank in *Enriching Lives* (1994) stated:

> The actual uptake of supplements by the targeted populations requires trained, motivated health care workers who can communicate effectively with consumers to overcome their fears, misinformation, and ignorance. (20)

> Thus, for targeted populations—and for mothers in particular, who must obtain supplements frequently, sometimes daily—merely showing up for the injection or actually taking the pill or giving it to a child often implies a great accomplishment....(health care workers must) explain the nature and importance of the capsules, pills, or injectables; to determine which family members need them and in what dosage and frequency; to tell when and where to get them; and to both warn and reassure the consumer about the supplement's possible side effects. (21)

Some studies have even documented that women will lie to researchers about whether or not they have taken the pills (Schultink et al. 1996).[20] Complaints about women abound. According to Viteri "the causes of failures of [supplementation programs]" were attributed to "mainly poor knowledge of the importance of adequate iron nutrition and anemia prevention, leading to late consultation and adherence to the supplementation regimen" (Viteri 1999, 17), and the failure of a supplementation program during pregnancy was ascribed to the "lack of compliance" (Lynch 2005, 334).

Experts see a similar problem with nutrition education and other related projects such as community and home gardening, which is theoretically another possible strategy to combat micronutrient deficiencies. For instance, in order to increase vitamin A intake, the consumption of dark green vegetables and fruits can be encouraged. For iron deficiency anemia, women can be educated about iron from animal and plant food. Since iron absorption is reduced by certain compounds in tea and coffee, women can be educated to reduce intake of them. Projects can help establish and manage community and home gardens that would grow diverse vegetables high in vitamins. Although nutrition education is almost always mentioned as part of "micronutrient strategies," it has never become the primary policy. In fact, the minority of nutrition experts who are in favor of such an approach lament that nutrition education and community-based projects only get lip service (Underwood and Smitasiri 1998). As we saw in chapter 2, experts' preference for quantifiable results might be one reason for this lack of enthusiasm. Nutritional education poses a significant difficulty in designing experiments and measuring the outcome of behavioral change. The former secretary of the UN's Administrative Committee on Coordination, Sub-Committee on Nutrition, Leslie Burgess, however, hints at another source of experts' ambivalence about nutrition education:

> I think it was the feeding programmes which were a major problem, on the other side nutrition education was and to some extent is, viewed with suspicion by the conventional medical practitioner. You could persuade Mrs. X to eat less fat, or that she has to breastfeed her kids instead of [using] a bottle; it's all loose stuff. Whereas someone has produced a relatively new antibiotic which zaps a particular bug. So if you're in third world medicine, it's a lot more comfortable to go along with nicely defined things. If Mrs. X does not feed her kid well, and the kid dies, you feel responsible. (quoted in Ruxin 1996, 334)

In addition to being "loose stuff," Burgess's comment describes experts' frustration with having to (and often failing to) persuade women to stick to nutritional advice. Elaborate nutritional education workshops and training sessions can be devised, but the problem of compliance always lingers in the minds of experts and bureaucrats. Similar frustration with recalcitrant "Mrs. X" in developing countries is implicit in a statement from the World Bank (1994) that highlights the drawbacks of the nutrition education approach: "Consumers must believe that the desired change in their dietary behavior will bring tangible benefits. Vitamin A programs in four Asian countries could not persuade mothers to give green, leafy vegetables to their young children to avoid blindness, a malady too rare to compel a change in behavior" (33). Implicitly identifying these women as an obstacle for nutritional improvement, critics of nutritional education underscore the fact that women tend not to follow the advice given in nutrition education and that it is not possible to be sure that women's cooking and eating habits change *and* stay changed once experts' monitoring stops.

In contrast to these more complicated and competing micronutrient strategies, fortification is "better" from the perspective of many experts in that there is little need to rely on women's collaboration to control the intake of micronutrients. The key word here is *control*. Without having to ask women about their eating habits or to educate them about nutrition or to convince them of the efficacy of the solution that is being offered, fortification can nevertheless increase the amount of micronutrients circulated in the bodies of people. With fortification, selected nutrients can be added to whatever people already eat—be it cookies, instant noodles, or condiments. Hence, to experts, fortification involves less uncertainty and a greater sense of control. Technically, manufacturing fortified food may be complicated, but nutrition experts tend to share basic vocabularies, assumptions, and commitments with manufacturers, whereas poor and uneducated Mrs. X may be difficult to work with. In an important way, fortification enables nutrition experts to *bypass women, rather than engaging with women* to tackle food problems. It's easier for these experts simply to work with experts from the industry side.

This erasure of women from the process is paradoxical, because in comparison with previous charismatic nutrients, the need for micronutrients seemingly involves greater gender consciousness. Women are often identified as victims of micronutrient deficiencies, as the FAO states in its "State of Food Insecurity in the World 2002": "Children and women are the most vulnerable to micronutrient deficiencies…women because of their higher iron requirements, especially during childbearing years and pregnancy" (FAO 2002b, 24). Pointing out that women bear "the heaviest toll from these dietary deficiencies" (Kennedy, Nantel, and Shetty 2003, 8), experts have rallied the world's support for micronutrient projects.

The relative prominence of women's needs is not accidental, as the 1990s saw an impressive amount of global development activities related to gender. Institutional integration of gender into international development was called for, and many international organizations responded by "mainstreaming gender" into all aspects of international development to rectify the previous separation and marginalization of gender issues.[21] People in charge of micronutrient projects often saw themselves as fitting into this larger trend, and they emphasized their consciousness of the necessity of empowering women in developing countries. Nutritional experts at the Asian Development Bank, for instance, touted the fit of micronutrient projects into the "gender mainstreaming" imperative by saying that "the educated and socioeconomically empowered Asian woman is the key to improving the nutrition and mental acuity of young children, and that such improvement sets in motion lifelong prospects for heightened learning and earning with benefit streams to families, communities, and nations.…*Mainstreaming gender concerns is essential* if nutrition programs are to succeed" (Hunt and Quibria 1999, iii; my emphasis).

Although salutary in its intent, gender mainstreaming has not been immune from criticism by feminist scholars. Some argue that international organizations often merely have introduced gender perspectives to existing policy programs without challenging the old paradigm. That is, gender issues have been added, but they have not led to a fundamental rethinking of mainstream approaches to development issues that could "reorient the nature of the mainstream" (Jahan 1995, 13). Or worse, some criticize that women are now to be used as the "agent-as-instrument of transnational capital's globalizing reach" (Spivak 1999, 201–2). Feminist radical politics are frequently sidelined vis-à-vis mainstream international development goals while projects that have higher resonance with mainstream international development get priority (Kabeer 1999; Jahan 1995).

The issue of gender has been brought into development policy as a means to an already established goal, not as a means to redefine the goal, or gender equality as a goal itself.

Contradictory gender projects also have emerged with micronutrients and fortification. On the one hand, it has brought "women" as a topic into the discussion of micronutrients, and women's micronutritional status has become a salient rallying cry for those who want to introduce fortification. At the same time, fortification schemes have tried to improve women's nutritional status by bypassing, rather than by engaging with, poor malnourished women themselves. The possibility of giving more power and autonomy to these women goes against nutritional experts' long-standing doubt about women's capacity to act in accordance with modern nutritional knowledge. Instead, uneducated poor women are often identified as the bottleneck of schemes to improve the nutrition of the global South.

Therefore, the featuring of women in discourses surrounding micronutrient deficiencies suggests much to be considered by feminist scholars. Of course, women menstruate, carry children until their birth, and do the most to feed them in early years, hence, their nutritional status affects children's as well. Yet we must consider the implications of women being marked as a "vulnerable" population in relation to micronutrient deficiencies. I suggest the concept of *biological victimhood* to understand the complicated visibility afforded to women in contemporary food policy discourses. As noted in chapter 1, biological victimhood refers to a delimiting perspective within the medical and food policy circuit that affords a space for women in food reform debates based on their biological propensity to a particular group of diseases, disorders, and disasters. It gives women visibility in food policy, but only as a biologically sexed group of likely "victims."

The visibility brought by biological victimhood is tricky, and echoes Wendy Brown's observation on the politics of "identity of injury" in which "the language of recognition becomes the language of unfreedom...a vehicle of subordination through individualization, normalization, and regulation, even as it strives to produce visibility and acceptance" (1995, 66). Although such a political move intends to recognize women's past injuries and embodied vulnerabilities, Brown worries that such projects, although well-intentioned, ironically reenact the very effects of power that they try to overcome.

Some feminist scholars researching international development have echoed Brown's concern, noting that the blanket identification of women as "vulnerable" can backfire and produce unintended consequences that ultimately hinder feminist ideals (Parpart 1995; Enarson and Meyreles 2004; Fulu 2007). They do not deny that women have certain vulnerabilities and historically and culturally constructed handicaps. Bringing visibility to gendered impacts of development interventions is a step in the right direction after the previous approach, which neglected gender as an important variable in policymaking. But a universalizing statement about women as biologically determined victims has serious political consequences.

Following these scholars, I argue that in the case of hidden hunger, several layers of this feminist paradox must be recognized. First, while visibility of women might be a welcome improvement from the previous neglect, the universalizing categorization of "women = vulnerable" obfuscates the fact that vulnerabilities are dependent on a complex interplay of factors, including gender, class, ethnicity, and age. We need to recognize that the different positionalities of each woman produce different vulnerabilities (Fulu 2007). In contrast, biological victimhood depends on the "abstract individuation" that has been criticized by feminist philosophers (Sprague 2005, 16–18). Biological victimhood's abstracted frame of analysis cannot adequately describe the everyday constraints and needs of women in facing the food problem because it does not account for complex intersectionality (Collins 1998).

Second, to notice that the visibility accorded to women under nutritionism assigns them to biological victimhood is important because of its political consequences; women are seen this way as the victims of their own biology but not the victims of politics or social positionality. Paradoxically, the universalizing categorization and overdetermination of women as victims throws the responsibility for poor health onto individuals by making their vulnerability essentially a biological one. For instance, women's iron deficiency anemia is understood to be a simple function of women's bodily functions, such as menstruation and pregnancy. Such a biology-centric analysis effectively masks the fact that vulnerability is also a derivative of political and social factors. How did these women become malnourished in the first place? Why are they deprived of means to address the problem? Social relations that critically constitute vulnerability become obfuscated.[22]

Furthermore, women's identification as vulnerable victims of micronutrient malnutrition has relegated them to the position of passive *recipients* of food aid and nutritional expertise. Such a depiction of women as passive and needy victims risks reinforcing a gender stereotype that portrays women as inherently weak and powerless. Scholars have observed that expert-driven, state-sanctioned interventions are made in the name of women's disadvantage and vulnerability, but actually they neglect women's capacities, resources, and longer-term interests (Fulu 2007; Clifton and Gell 2002). The governments and experts become the active, benevolent "doers" of things, as they further reach into the lives and bodies of women.

Ultimately, biological victimhood in tandem with nutritionism brings women into food politics not as individuals embedded in the context of social relations with differing needs and priorities but as abstracted members of a biological group with inherent nutritional disadvantages. This then allows experts to take charge of defining needs and wants rather than compelling them to give opportunities to women to decide for themselves. Women's experiences and hopes for future directions are presented as already known by science—the first is their suboptimal nutritional status and so the second must be additional nutrients. Without obvious symptoms discernible to lay people's eyes, "hidden hunger" is the quintessential example of such a naturalized imposition of victimhood. The paradox of the visibility of women under nutritionism is that attention to the victimhood of women does not lead to an imperative to listen to them, because women's victimhood is seen in terms of biology, so their "needs" are clearly known by scientists, even better than by themselves.[23]

It is precisely this lack of space for women to determine their own problems and prescriptions that Wendy Brown highlights when she differentiates the "problem of the good" from the "problem of the true" (1995, 49). Brown points out that while experts might assume that they know the "truth" about women (such as their "nutritional status"), the real politics should be about what women want for themselves. Women need a space "for discussing the nature of 'the good' for women" (49) rather than having it dictated by experts. Indeed, if we think about vulnerability in broader terms, an often neglected feature of vulnerability is a lack of participation and involvement in decision making and policy processes. Yet the prevalent "women = biologically vulnerable" equation ironically blinds us to this key social dimension of vulnerability.

Although useful in making women visible in food policymaking, the universalizing identification of biological victimhood hinders a more feminist strategy for increasing women's participation in food policymaking under conditions other than dependency. Nutritionism's paradox lies in the obfuscation of women's capacity to construct their own understandings of food and nutrition, while purporting to bring women to the much-deserved attention of international development experts.

4

BOUND BY THE GLOBAL AND NATIONAL: INDONESIA'S CHANGING FOOD POLICIES

**Food and Nutrition for the Future: Increasing Productivity
and Competitiveness of the Nation (Pangan dan Gizi Masa
Depan: Meningkatkan Produktifitas dan Daya Saing Bangsa)**

—National Workshop on Food and Nutrition (Widyakarya Nasional
Pangan dan Gizi) theme, 1998

**At the launching of the Healthy and Productive Female Worker
Movement at the vice president's palace, vice president Try Sutrisno
said that women workers need to increase work productivity,
although their primary duty was to educate children and deal with
household issues as housewives.**

—*Suara Pembaruan,* November 14, 1996

My stories of charismatic nutrients have focused on the international stage, touching on varying roles played by organizations and scientific experts. Now I turn to the question of how charismatic nutrients become local. The following four chapters anchor the global stories of micronutrients in a local setting: overall food policy (chapter 4), mandatory fortification (chapter 5), voluntary fortification (chapter 6), and biofortification (chapter 7) in the context of Indonesia. With relatively recent exposure to fortification and biofortification, Indonesia offers a suitable site for analyzing their dynamics. It would be easy to naturalize the growing influence of micronutrients in Indonesia. The country has achieved impressive economic growth and thus perhaps a focus on quantity is no longer necessary. However, I will scrutinize such naturalized assumptions about changes in food policy and point to the dynamic configurations of the diagnosis of, and solutions to, the food problem.

The stories of charismatic nutrients and the global turn to micronutrients bring up questions about their local translations: How did the global shift toward micronutrients in food policy play out in developing countries? How do charismatic nutrients and nutritional fixes travel from the international realm to a developing country? Although "global" stories tend to be told as if they have automatic global reach, in reality the global articulates with the local in complex ways rather than in a linear, top-down fashion. The ontology of the "global" and "local" is also not so simple, and we need to complicate their social constructions as well. Rather than seeing the travel of micronutrients as a "norm diffusion" from the global center to the periphery, I draw on theories of "biopower" and its transnational mobilizations. As feminist theorist, Nancy Fraser points out, biopower is unbounded by national borders (2009). Fraser's concept of "globalized governmentality" (125), and a similar concept of "transnational governmentality" by Ferguson and Gupta (2002, 981), point to the need for examining "a new multi-layered regulatory apparatus, which operates on a transnational scale" (Fraser 2009, 126).

Biopower that spills over national boundaries can be seen in the case of Indonesia in its overall food policy. On the one hand, the micronutrient status of Indonesians increasingly became of international concern. At the same time, Indonesians themselves also figured into the picture. However, the micronutrient turn in Indonesia was not simply a reflection of international consensus—calculations by local actors also mattered significantly. In this chapter I look at both international actors such as UN agencies, bilateral aid agencies, and international NGOs and the Indonesian domestic players. The increasing charisma of micronutrients in the 1990s in Indonesia was a product of such "global assemblages" (Ong and Collier 2004, 1). Global assemblages of players supporting micronutrients, however, had a particular structure. I began to realize this when I participated in a workshop on fortification in Jakarta in December 2004. An influential nutrition researcher told me about an "important" meeting on food fortification. I asked whether I could attend, and he kindly coordinated my participation. The meeting was held in a hotel in Bogor, a favorite getaway for urbanites from busy Jakarta. Entitled Workshop on the National Plan of Action on Food Fortification, the two-day meeting explored the direction of national policies on fortification. Interspersed with coffee breaks and buffet meals, the meeting consisted of several presentations ranging from "Principles of Strategic Plan of Action in Management" by the industry representative to "Suggested Strategy, Objective, Target, Output, and Outcome of the Future Fortification Program for Developing the National Food Fortification Program" by university researchers. The discussion primarily centered around technical issues: what to fortify (oil? sugar?) and how to collaborate with the food industry to promote fortification (cost? marketing?).

After watching numerous PowerPoint presentations—one was by a speaker from a flour mill industry who enthusiastically touted business-type strategic planning by quoting ancient Chinese war strategists—I felt that malnutrition and hunger in the country were strangely distant and abstract. Indeed, the meeting was indicative of the global assemblage of micronutrient advocates: there were forty-five "experts" from government, universities, international organizations, and the food industry. While this group had both international and national representatives, glaringly absent were ordinary Indonesians who could share their stories of the cost of feeding their families and agonizing over ill children. The workshop, which was attended only by experts, domestic officials, and industry representatives, reflected the two powerful forces that shaped the micronutrient turn in Indonesia: the international policy trend and national development priority.

The micronutrient turn in Indonesia is a part of the transnational regime of truth production and discipline that facilitated new scientific and social logics for interpreting the state of health and nutrition in the country. What are the driving forces behind the rise of micronutrients in Indonesia, and who was made invisible in the policy debate? International as well as national priorities set the stage for the promotion of charismatic micronutrients in Indonesia, but *whose* priorities were they? In looking at, in particular, the micronutrient project for women workers, I question the construction of priorities that privilege productivity over justice.

Charismatic Micronutrients: The Indonesian Story

In the 1990s, micronutrients came to the fore of nutrition policy in Indonesia. First, the government widened vitamin A capsule distribution. The earlier program of vitamin A capsule distribution targeted only children aged twelve to fifty-nine months, but in 1991 the government started to target pregnant women as well (de Pee et al. 1998). And in the late 1990s, the government further expanded the program to include postpartum mothers and infants of six to twelve months as recipients of capsules (Soekirman et al. 2005; Helen Keller International 2000). The government also drew resources from international organizations to conduct a variety of related research projects in the 1990s. For instance, it acquired funding from Helen Keller International and USAID to conduct several vitamin A promotion projects nationwide (Pollard and Favin 1997) and from UNICEF for a similar vitamin A project in Central Java in the 1990s (de Pee et al. 1998). The government and HKI collaborated on a project called ROVITA, an oral rehydration and vitamin A project that promoted vitamin A capsules and oral rehydration therapy among an additional 23,000 children in

Central Java. They also conducted social-marketing campaigns for vitamin A, promoting vitamin A capsules and vitamin A–rich foods among 40,000 children in one district in West Sumatra (Shaw and Green 1996; Soekirman et al. 2005).

In response to iodine deficiency disorder, experts replaced an earlier government program of injection with iodine capsule supplements in 1992 (Soekirman et al. 2005; Direktorat Bina Gizi Masyarakat 1994). In 1993, iodine capsules for people in twenty-six provinces were prepared, and social-marketing campaigns using TV, radio, and posters were conducted (Direktorat Bina Gizi Masyarakat 1994). The government also renewed its iodine fortification program and experimented with the iodization of water in four provinces (Direktorat Bina Gizi Masyarakat 1997) and with salt iodization, mandating the latter in 1994 via a presidential decree (Sunawang, Lusiani, and Schofellen 2000). Experts received funding from international donors such as PAMM, UNICEF, and CIDA for salt iodization (CIDA 2006).[1] One of the bigger grants came from the World Bank for accelerating salt iodization from 1996 to 2003 (Soekirman et al. 2005; Sunawang et al. 2000).

The government also accelerated efforts to reduce iron deficiency anemia. The IDA program in the 1970s targeted pregnant women, providing iron capsules every day for ninety days during pregnancy and for forty-two days during the postpartum period. But implementation was poor and the proportion of pregnant women actually taking these capsules was quite low. Since the late 1980s, the government had introduced many measures to increase this rate, such as increased supply and availability of supplements at each level of the health system, improved packaging, social-marketing campaigns, enhanced availability of program guidelines and protocols, and monitoring systems for anemia and supplement use.

In addition, the IDA program expanded its target population to include women of child-bearing age, "brides-to-be," and teenage schoolgirls, encouraging them to take iron tablets regularly once a week (Kurniawan 2002). Experts were able to get endorsement not only from the Ministry of Health, but also from the National Family Planning Board (BKKBN), the Ministry of Education and Culture, the Ministry of Religious Affairs, and the Ministry of Social Affairs (Soekirman et al. 2005). Furthermore, in the late 1980s, the government started programs to specifically address female workers' IDA, and in 1992, the project received renewed emphasis. The Ministry of Health, the Ministry of Manpower, the BKKBN, the State Ministry of Women Empowerment, the Ministry of Education and Culture, and the National Development Planning Board (BAPPE-NAS) started a long-term anti-anemia strategy for female workers. In 1996, the Ministry of Health and the Ministry of Manpower issued a decree on reducing anemia in female workers (Direktorat Bina Gizi Masyarakat 1996).[2]

Similar to the global embrace of fortification, Indonesia also saw increasing official commitment to fortification. Fortification was first mentioned in the nation's fifth five-year development plan in 1989 (Repelita V, 1989–93),[3] and the government finally decided on salt iodization and wheat flour fortification in the late 1990s. Wheat flour fortification, which was one of the programs to combat IDA, is discussed in detail in chapter 5. A fortified baby food program was also begun at this time, which is discussed in chapter 6.

The government's and experts' data-collecting activities relating to the food problem also indicate the growth of interest in the micronutrient status of the population in the 1990s. It may seem easy to gauge a nation's nutritional situation, but in actuality it is no simple task. For a long time, the Indonesian government relied on data on food production (availability of protein and calories) rather than on the nutritional status of the population per se. The nutritional situation was estimated based on a "food balance sheet" (*neraca bahan makanan*), which was a set of data comprising domestic food production, exports and imports, availability, food loss, as well as human consumption (Arifin 1993, 163). Another frequently used way to infer the nutritional status of the country came from a household food intake survey conducted as part of the National Social Economic Survey (SUSENAS). It asked respondents to recall what they ate in order to measure food intake by households. This survey was started in 1963. In the 1980s, the government sought funding from USAID to add child anthropometry data with the hope that this would be a more direct measurement of national nutritional status. A national survey called the Integrated Nutrition Survey started to integrate measurement of the weight of children under five years of age into SUSENAS (Surbakti 1987; 1994). The data collected by these methods was perhaps useful for assessing the *macro*nutrient condition of the population, but it could not estimate reliably the micronutrient situation. The national prevalence of micronutrient deficiency was long unknown. Estimates of the prevalence of vitamin A deficiency, iron deficiency anemia, and iodine deficiency disorder were not available, and it was not until the 1990s that data gathering intensified. For VAD, there was one national survey in 1977 called the Nutritional Blindness Survey, conducted in collaboration with Helen Keller International. But this survey focused on xerophthalmia rather than VAD in general. In 1992, at the urging of scientists and health bureaucrats who wanted to know the status of VAD in the country, the government conducted another survey, the National Xerophthalmia Survey. For IDA, the government conducted a survey in 1986 as part of the National Household Health Survey (Survei Kesehatan Rumah Tangga or SKRT), which provided the first national data on hemoglobin levels of pregnant mothers. The government tried to institutionalize this IDA assessment, and so the two following SKRTs, in 1992 and 2001, measured hemoglobin levels of pregnant mothers, reproductive-age women, and children under five years old. For IDD, the National Goiter Survey was conducted once in 1980, but there was no follow-up for a long time. In the 1990s, in response to the renewed attention to IDD, the government conducted a series of national surveys in 1990, 1996, 1998, and 2003 (Azwar 2004). In short, national data on food availability and child weight existed from the 1960s, but it was only in the 1990s that data on micronutrient deficiencies began to be collected with any regularity. Before that, national data on vitamin A deficiency, iodine deficiency disorder, and iron deficiency anemia was quite limited.[4]

Along with data, another key development was in the use of the term "micronutrient" itself. Although Indonesians had translated English words such as "vitamin" and "protein" into Bahasa Indonesia, the official language of Indonesia, the word, "micronutrients" did not have an Indonesian counterpart until the early 1990s. In my interviews with Indonesian experts, it emerged that the term's translation was first discussed in bureaucratic meetings in 1993, when the government nutritional experts were debating the nutrition policy for the coming five-year plan (Repelita). Many Indonesian nutrition experts, both at universities and governmental agencies, had been educated in the West and were aware of the global turn toward micronutrients. They realized the need for an Indonesian word for the concept, and after some discussion, they agreed to the translation *gizi mikro*, which literally means "micro" (*mikro*) "nutrient(s)" (*gizi*). Some thought that this phrase might mistakenly give the impression that these nutrients were unimportant because *mikro* connotes something small.[5] Nonetheless, *gizi mikro* became widely accepted in the lexicon of Indonesian nutritional science.

This new term, *gizi mikro*, has had an interesting social function by providing a new category that has reconfigured and extended technoscience networks. Researchers started to identify themselves as doing analysis on *gizi mikro* instead of saying, for example, that they do research on vitamin A or iodine. The term *gizi mikro* also engendered a bureaucratic reorganization. The Ministry of Health decided to create divisions of micronutrients and macronutrients (Gizi Mikro and Gizi Makro) under the Directorate of Community Health.[6] This process facilitated communication with international actors, who then shared an identity as micronutrient researchers. It also created a space for Indonesian researchers who were empowered by the growing global charisma of micronutrients and were capable of speaking on behalf of the related global consensus.

With expanded policy programs, improved data sets, and the lexical entry, micronutrients began to figure centrally in Indonesian food policy in the 1990s.

Paralleling the global trend that we have seen in earlier chapters, micronutrients came to exert charisma in the Indonesian food policy community, attracting funding, modifying the institutional configuration of bureaucracy and scientists, and shaping policy interventions.

Synchronizing Food Policies

How do we explain this Indonesian turn to micronutrients? It would be difficult to simply say that Indonesia "uncovered" hidden hunger in the 1990s. There had been multiple studies in Indonesia, albeit on a small scale, that indicated the prevalence of nutrient deficiencies since the 1960s (Martoatmodjo et al. 1972, 1973, 1980; Karyadi 1973a; Permaesih, Dahro, and Riyadi 1988; Dahro et al. 1991). It would similarly be difficult to attribute the shift to the eradication of problems of macronutrition or protein-calorie-malnutrition. Rice self-sufficiency was achieved in 1984, but it has not been maintained. Indonesia still faces the problem of lack of food and low-caloric intake. Scientific and technological advancement alone cannot explain the charisma of micronutrients at a particular historic point, and social factors ought to be considered.

From the above description, it is undeniable that international organizations played a significant role. From the World Bank's iodine project to UNICEF's vitamin A project, many of the micronutrient projects in Indonesia were funded or prompted by UNICEF, USAID, WHO, and other international organizations. Nongovernmental organizations based in the United States, such as Helen Keller International, also have played a critical role in directing more resources to micronutrient-related projects. As we have seen, there have been many international agreements that have aspired to tackle global hidden hunger, and they have required local sites and willing collaborators to realize their claim of having a global reach. Project financing, training, workshops, conferences, and pilot studies are all part of an important path through which the global discourse finds concrete points of engagement.

In addition, international organizations set up local counterparts to the international initiatives, which facilitate the global to local translation. For instance, the local Indonesia Fortification Coalition (Koalisi Fortifikasi Indonesia or KFI) backed fortification in Indonesia. The KFI includes many influential Indonesian experts in nutrition and food technology, government officials from the Ministry of Health and the Ministry of Trade and Industry, and business interests such as the Chamber of Commerce's Division of Food and Beverages.[7] Its genealogy is telling of the influence of international organizations. In 2000, the Asian

Development Bank organized a conference called the Manila Forum to promote fortification in Southeast Asia (ADB 2000b). It was on the recommendation of the Manila Forum, and with the funding from UNICEF, that the KFI was established in 2002. The KFI subsequently has served as a nongovernmental local liaison for international donors that are interested in promoting micronutrient projects in Indonesia.

The presence of cosmopolitan Indonesian food and nutrition experts also has eased the journey of micronutrients to Indonesia. Local groups such as the KFI tend to include Indonesians who are fluent in English, many of whom have academic degrees from American or European universities. From the perspective of international organizations, they are easy to communicate with, not only because of their fluency in English but because they share the same kind of "development" language and an understanding of global trends, including the trend toward micronutrients. These Indonesian experts are well aware of current beliefs and practices in international nutrition and nutrition-related development programs around the world. There is little need to preach to them about the importance of micronutrients or the seriousness of micronutrient deficiencies.

Global-local interactions are not clear-cut. Indonesian experts and organizations embody "the local" vis-à-vis the "global," and having such local partners is important in that local consultation, participation, and collaboration is valued in international development. These local experts are not merely transmitters of global norms, a passive node through which the global "epistemic community" channels its consensus after it is already formed—as understood in world society theory. Rather, they are a kind of hybrid group that also participates in the formation of the global consensus and trends. Many nutritional studies on micronutrients were conducted in developing nations, with Indonesia being one. The most influential study of vitamin A, the Aceh study, by Alfred Sommer (discussed in chapter 2) took place in Indonesia. Dutch nutrition researcher Saskia de Pee and her colleagues (1998) conducted many vitamin A–related projects in Indonesia as well. These foreign researchers needed government approval for conducting research, institutional sponsors in Indonesia who would agree to write letters to relevant agencies, translators, and other local staff to coordinate the projects' logistics. Many of the cosmopolitan Indonesian researchers also participated in international nutritional organizations that have influenced the direction of global discourse. Muhilal and Darwin Karyadi,[8] both of whom have led Indonesia's most prestigious nutrition research center, the Center for Research and Development of Nutrition and Food (Puslitbang Gizi), have published in Western nutrition journals. Many of the Ministry of Health's nutrition experts studied abroad and were collaborators in Western researchers' nutrition studies.

Soekirman, who was the president of the Indonesian Nutritionist Association (Persatuan Ahli Gizi Indonesia or Persagi), perhaps best embodies this hybridity. He studied initially at the Nutrition Academy in Indonesia and then attended Cornell University in the United States for his graduate degrees in international nutrition. He served in many influential organizations in Indonesia, most notably at the National Development Planning Board. Simultaneously, he has been active in international nutritional circles. When I interviewed him in his office in Jakarta, which was littered with policy reports from international organizations, he said of his achievements: "I was president of nutrition societies. Not only in Indonesia, but also in Asia and internationally. So I have been very much linked with the world. At any international nutrition society, they know me. I am a director of ILSI Southeast Asia. I was an expert adviser to the UN Sub-Committee on Nutrition in Geneva. I attended every annual meeting. So we are very close to international scientific groups." Rather than being a passive conduit of an externally formed global consensus, Indonesian experts like Soekirman are part of the global force.

Furthermore, while international donors have sought to implement projects in Indonesia, we should not consider Indonesians as merely manipulated by international actors. Indonesian experts have been willing collaborators in international endeavors. International projects, research collaboration, conferences, and workshops are important sources of funding and prestige for Indonesian researchers and bureaucrats. They are acutely aware of the need to be attuned to international discursive changes so as not to miss new opportunities for funding and prestige for themselves and for their organizations. The ebb and flow of foreign aid, whether bilateral or multilateral, public or nongovernmental, is ingrained in their lives and careers. Development projects tend to move from one theme to another, making different issues the poster child at different times. In one year it might be "democratization," while in the next it might be "civil society." For people dealing with international organizations, the pressure to navigate and adjust well to these changes in donor preferences is nothing new. They would rather harness this flow of change than become victims of it. If international organizations have thought that they were persuading these locals to take up the next big thing in development, Indonesian experts have been similarly astute in being persuaded and receiving funding, international travel, and prestige.

Ana Tsing has argued that "global forces are themselves congeries of local/global interaction" (2004, 3). International trends in nutrition have found Indonesian expression through such congeries of the local and global. Through the network of international organizations and their Indonesian collaborators, the importance of micronutrients has been disseminated throughout Indonesia. These local and global actors, with differing but overlapping commitments and ambitions, have shaped the direction of Indonesian food policy.

Resonance with a Development Paradigm

In the context of modern developing countries, the nation's relationship to tech-noscience is shaped by its project of nation building, and in particular the mandate for the state to "develop" itself. From the atomic bomb (Abraham 1998) to nuclear power (Hecht 2000) to the megatelescope (Abraham 2000) projects, science and technology represent the "epitome of and metaphor for the modern" (Abraham 1998), constituting what "underdeveloped" countries should strive for. The quest for technoscience capability is thus embedded in nation states' understanding of modernization and development. Nutritional science is part of this. Nutritional science has also provided a useful instrument for states as a basis for welfare intervention (Kjaernes 1995), driven by the desire to secure a cheap and healthy labor force (Aronson 1982; Turner 1982) and by military needs to produce healthy soldiers (Burnett 1979; Levenstein 1993). Policies on nutrition, food, and bodies are not merely to be considered as humanitarian in their intentions. Rather, as in other policies, they aid certain types of social engineering by the modern state. An inseparable part of food policy is the state's logic in pursuing a particular shape of citizenry and nationhood. Therefore, to consider Indonesia's micronutrient turn in relation to the global is to tell only half of the story. Besides the alignment with global trends as described, Indonesian food policy also needed to fit with aspirations of the Indonesian state. Management of bodies of the nation is an integral dimension of development, and to understand the power of micronutrients in Indonesia, we need to examine the relation between national development priorities and food policy.

I have discussed how the charisma of micronutrients, and in particular its accompanying nutritional fix, fortification, has had a strong resonance with neo-liberalism. This can be seen in the context of Indonesia. The important political context of the micronutrient turn was the shift in overall development policy from state centered to market based in the 1990s. Far from static, dominant thinking on what a nation has to do to "develop" and what "development" means is subject to constant change. Indonesian economist Thie Kian Wie provides a useful overview of shifting development paradigms in Indonesia. He observes that from the 1960s to the mid-1970s, Indonesia's national priority was recovery from the economic turmoil caused by political upheaval. The period after 1974 was shaped by a major oil boom that fuelled rapid economic growth. Development planning became centered on building infrastructure and nurturing domestic industry by import substitution and a protectionist trade policy. When the oil boom went bust in the 1980s, the government changed gears and turned to deregulation, liberalization, and an export-led growth model of economic development (Wie 2002). In other words, since the 1980s, Indonesia increasingly took a neoliberal model of national development.

With this neoliberal shift, the previously dominant food policy programs—increasing food production via agricultural modernization and reducing mouths to feed via population control—came to be considered cumbersome and antiquated as they relied on state subsidies and were based on top-down bureaucratic structures. The necessary transition from such a state-heavy approach was envisioned as "from Green Revolution to Market Revolution," by the head of the Office of State Minister of Food Affairs and BULOG (Hasan 1993, 16).

Under the "market revolution," corollary changes in social policies were also necessitated, and accompanying the neoliberalization in economic policy was the new mantra of "human resource development." The improvement of human resources to create a competitive labor force was officially endorsed in Indonesia's second long-term development plan (Pembangunan Jangka Panjang II, or PJP II, 1994–2019) as the national goal (Ministry of Education and Culture n.d.). People now were the essence of the nation's survival in the global marketplace, and the prosperity of the New Order regime was to be built on an able human resource pool.

In the field of nutrition, this new economics of people brought the economization of nutrition that was discussed in chapter 3. Mirroring the international situation, economization of nutrition has also compelled Indonesian food and nutrition experts to use economic frameworks in diagnosis and prognosis of food problems. No discipline could afford to be irrelevant to national development, and nutrition experts refined their framing of nutrition and the food problem to fit the emerging view of nutrition. Consequently, the traditional etiological emphasis gave way to a productivity focus. Many iterations can be drawn between nutrition and the new development priority (see fig. 4.1). The key message is that nutrition contributes to human resource development and hence ought to be considered as an investment.

It was in this context of economization of nutrition that micronutrients emerged as a key link between nutrition and national development. This is not to say that there was a natural fit between micronutrients and the new development priority, however. Rather, experts subtly shifted their framing of micronutrient deficiencies from being a survival issue to a competitiveness issue, highlighting their impacts on cognitive functioning, work capacity, and productivity. We

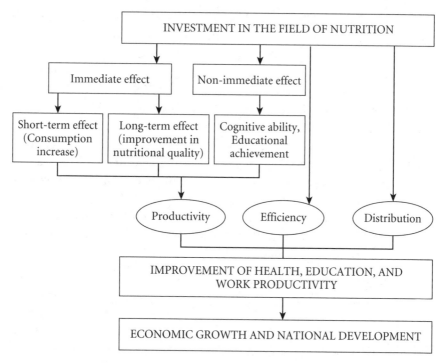

FIGURE 4.1. Illustration of the link between nutrition and development.
Source: Jalal and Atmojo 1998, 923; my translation.

can see this shift in the new representation of various micronutrient issues. For instance, iron deficiency anemia, departing from the previous representation as a particular medical condition (anemia), received renewed attention in the 1990s as a work-productivity problem (WHO 2001b). Often cited as the justifications for IDA prevention in the 1990s were impacts on labor performance, including a study on road construction workers that found that anemia was closely associated with poor work performance (Basta et al. 1979; Karyadi 1973b; Husaini, Karyadi, and Gunadi 1981). Refashioned as a matter of productivity, IDA became a problem of loss to the national economy.

Similarly, experts were able to renew attention to iodine deficiency disorder in the 1990s by linking it with mental and intellectual impairment. In the earlier period, IDD was called "endemic goiter" rather than "iodine deficiency disorder." That is, the consequence of the deficiency was seen as the swelling of the thyroid gland (goiter). Now, recategorized as IDD, emphasis became on its influence on intellectual ability. It was only in Repelita V (1989–93) that the term "iodine deficiency disorder" was used with its emphasis on mental impairment rather than goiter. Placing IDD differently, Repelita V started to construct and justify the importance of tackling the IDD problem as one of protecting children's intellectual ability, and Repelita VI continued this theme.[9] Compelled by the new emphasis on human resource productivity, interpretation and representation of micronutrient deficiencies became significantly modified.

In addition to the contribution to human resource development, experts framed micronutrient deficiencies as fitting the doctrine of less government. Fortification was constructed as the most obvious market-friendly policy in the nutritional field. Fortification satisfied the two-pronged demands of the new development paradigm: to achieve human resource development by using a market mechanism. The link between fortification and Indonesia's new national priority was made clear in the words of the primary fortification promoter from the Office of State Minister of Food Affairs in 1997, who explained fortification as follows:

> The meeting of economic leaders of APEC in Bogor in 1994 decided that trade and investment in the region needs to be liberalized by 2010 for developed countries and by 2020 for other APEC countries. This liberalization cannot be avoided, especially after ratification of WTO in 1995. For the implementation of WTO agreement the most important is human resources. Even now, the impact of liberalization is increasingly felt.... Moreover, AFTA implementation (in 2003) is only six years from now. Because of that, the meeting on Food Product Fortification to Improve Human Resource Development is very important. The issue of human resources is a priority goal in PJP II. (Natakusuma 1997, 1; my translation)

Note how Natakusuma links economic globalization to the necessity of human resource development and to fortification. At the National Workshop on Food and Nutrition (Widyakarya Nasional Pangan dan Gizi), he delivered a speech entitled "Food Fortification Strategy," in which he underscored the link between fortification, human resource development, and national development, saying that "in the current competitive age, quality of human resources is the key factor for development." In the eyes of the state with its neoliberal development plan, human resource development was the first layer of justification for fortification; that it could be achieved by market mechanisms was an additional appeal.[10]

The parallel between development paradigms and food and nutrition policy that I have described is not only a matter of experts' strategic positioning vis-à-vis the state development paradigm. They are also severely bound by institutional mechanisms to strictly synchronize their activities with the development paradigm. In the daily conduct of bureaucratic administration and academic research, nutritional experts have had to act within the confines of overall development goals. In Indonesia, national development was hierarchically structured with the Repelita on top, designed to align activities of the government at all levels. Repelita was a broad policy framework made every five years, which established goals and objectives for national development in the given period. After its approval by the parliament and the president, governmental agencies were asked to formulate programs and the accompanying budget in line with Repelita. The agencies' programs then had to be approved by the powerful National Development Planning Board, which made sure that these programs fit the overall theme of Repelita. Stipulating priorities of the nation in order to effectively mobilize its available resources for a unified goal, it set directions for sectoral activities. Changes in the national development paradigm were meant to be transmitted to the health, food, and nutritional sectors.

Academic research activities also had to align themselves with national development goals. The contents of Repelita and the budget approval process worked as a strong force that shapes the agenda for research in Indonesia. For instance, the central site of nutritional research in the country is the Center for Research and Development of Nutrition and Food under the Ministry of Health,[11] and the consistency of the Center's research agenda with Repelita was ensured by its internal review team and a review process at the Ministry of Health.[12] Research that did not have direct policy contributions was discouraged. The personal promotion of scientists at the Center was also tied to their contributions to the goals of Repelita. At the university level, too, the contribution of research to Repelita's goals was key to the survival of researchers in terms of the availability of funding. For instance, two major funding sources for nutritional scientists—Riset Unggulan Terpadu (Integrated Research of Excellence) and Hibah Bersaing (Competitive Grants)—evaluated research proposals on their relevance to Repelita's objectives.[13]

The Widyakarya Nasional Pangan dan Gizi has been another way that the development-nutrition linkage has been generated and ascertained. It is a meeting held in conjunction with the planning phase of every Repelita. Since its inception in 1969, Widyakarya has become an important event for showcasing the alignment of food and nutrition policy with overall development goals (Soekirman et al. 2003). To emphasize the meeting's importance, the president and ministers make appearances and typically give speeches about the importance of national development. Conceived as the expert body that gives input to Repelita, Widyakarya is tasked with creating a synergy between national development and food and nutrition sectors.

Nutritional and food policies and research agendas are tightly controlled to fit with the overall direction of national development. At the same time, nutritional scientists elucidate aspects of the food problem that match development mandates, since the alignment with national development is the key to securing funding, legitimizing one's discipline, and protecting institutional survival. Fashioning their programs as market-based ways improve the nation's "human resources," proponents of micronutrients aligned themselves with the development paradigm of the 1990s.

Women Workers as Resources

In November 1996, the vice president established the Healthy and Productive Female Worker Movement (Gerakan Pekerja Wanita Sehat dan Produktif).[14] This was a national campaign to encourage companies to distribute iron folate pills to female factory workers to reduce IDA (Kurniawan 2002). The government published the "Guide for Fulfilling Workers' Nutrition" (*Pedoman Kecukupan Gizi Bagi Tenaga Kerja*) to help improve food for workers at offices (Direktorat Bina Gizi Masyarakat 1997), took blood samples from female workers and mandated that companies give them iron supplements once a week for sixteen weeks per year (Kosen et al. 1998).

A seemingly innocuous public health program, this "movement" nonetheless reveals calculations on women's health within the economized logic. The movement was nominally a public health campaign, but it was simultaneously motivated by the need to cultivate a productive and efficient labor force for the purpose of national development. In fact, this was clearly articulated in the goal of the movement; according to the government, the movement's goal was "to increase awareness of the owners and managers of companies and developers to increase the health, nutrition, and productivity status of female workers in the framework of competing in the globalization era" (Direktorat Bina Gizi Masyarakat 1997, 46).

The government's seemingly benevolent concern for women's health needs to be juxtaposed with its repressive labor policy. The Indonesian government strictly managed labor under the doctrine of Pancasila Industrial Relations, which dismissed labor disputes as culturally unsuitable to Indonesia, and through a state-sanctioned labor federation called All-Indonesia Workers Union (Serikat Pekerja Seluruh Indonesia). Most workers were unable to organize or bargain, and workers risked heavy repression by the military and security forces if they tried to do so (Hadiz 2000). When Indonesia liberalized its business environment to attract foreign investment in the 1980s, the government wooed global capital with the promise of cheap and well-controlled labor (Spar 1996). Docile workers were crucial for national development.

The paradox comes into sharp focus with female labor. Women workers became the backbone of the national economy as Indonesia came to depend on export-oriented manufacturing after the bust of the oil boom in the 1980s. Replacing the petroleum industry as the core of Indonesia's economy were labor-intensive industries such as textiles, garments, and footwear, 80 percent of whose workers were women (Hadiz 2000; Tjandraningsih 2000). For many emerging Asian countries, female labor was important as the basis for export industrialization, but Indonesian women were positioned at the bottom of the regional economic hierarchy, providing cheap labor to neighboring countries as domestics or factory workers run by Korean or local subcontractors (Ong 2011).

Working conditions for these Indonesian female workers were abysmal. They were among the lowest paid workers in Asia. The Indonesian legal minimum wage was two dollars a day in the 1990s, which was not enough to cover basic needs (and furthermore was sporadically enforced) (Spar 1996).[15] Female workers typically received wages even lower than their male counterparts. Living and working conditions were also poor. For instance, a survey of female factory workers found high rates of intestinal parasites indicative of an unsanitary environment (White 1990). Moreover, these workers suffered from strict surveillance. Ong reported in 2000 that control and surveillance of female workers was found in "the provision of food, in granting or withholding of permission for menstrual leave, in the pressure for family planning and in physical confinement imposed during work hours," as well as "timing visits to the toilet, and using the excuse of having to verify requests for menstrual leaves to conduct body searches" (63).

The repression against women workers was most emblematic in the murder of a woman labor activist, Marsinah, in 1993. She was a twenty-five-year-old factory worker and labor activist in East Java who was tortured, raped, and murdered. The government blamed the factory management and arrested a few people, but they were found not guilty after a sham trial. An independent investigation by the Indonesian Legal Aid Foundation found strong evidence that the military was involved in the murder. As Rachel Silvey observes, the military "intended to terrorize women workers and discourage them from participating in labor activism" (2003, 138).[16]

As I have shown, micronutrients were wrapped around the concept of human resource development, but the story of the Healthy and Productive Female Worker Movement reveals underlying biopolitical calculations that shaped its interpretation. In theory, the concept of human resource development did not limit itself to a narrow range of issues that had direct economic return but had a theoretical breadth that included political participation and elimination of poverty and illiteracy. Yet the government's interpretation avoided dealing with any structural injustices. Underdevelopment, not injustice, was the problem for the government. The "human resources" of the country were to be *developed* (and exploited) but not *protected* for their basic rights, such as the right to organize, to bargain, and to have a livable minimum wage. Micronutrients provided a depoliticized window of opportunity, avoiding the possible radicalization of politics. In other words, micronutritional interventions helped the government to manage "populations in relation to the demands of world markets" (Ong 2002, 235) rather than managing the market in relation to the demands of the workers. As part of investments in the name of people's health and in the interest of global and national capital, micronutrient projects helped exacerbate the biopolitical control over the bodies of Indonesians.

Health without Justice

I have explored the complicated formation of political and social alliances that has propelled the charisma of micronutrients in Indonesia and have pointed out two "translations" that are important in this network of alliances. First is the global to local translation. There was global hype about "hidden hunger," and there emerged a network of experts and organizations that worked to translate its mandate into local policy in Indonesia. It was not an easy task, as the "global consensus" does not have automatic power to cause sweeping change in a given locale. International organizations and NGOs cultivated links with Indonesian experts and bureaucrats through involving them in research projects, establishing counterpart NGOs, putting on workshops and conferences, and through project financing. Simultaneously, the engagement with the "global" was also sought by Indonesians. Attuned to the ebb and flow of global development discourses, many Indonesian experts were eager to take part in this micronutrient turn, which they rightly saw as a way to establish or strengthen their connection with the cosmopolitan world and to increase resources that were so lacking within the country.

In addition to the global to local translation, another level of translation had to be performed. This was a translation of the meaning of micronutrients and micronutrient deficiencies to fit better with the development discourse of the day. Being a part of "national development" is important for the nutritional field, given its historic marginalization. Population control and agricultural sectors had occupied the top of the national priority list, drawing strong political commitments and government resources. To claim its space in the development apparatus, the nutritional sector had to construct itself as contributing to

national aspirations. Therefore, translation of the social meaning of micronutrients to fit with the contemporary development paradigm was an important
institutional and personal investment for Indonesian nutrition experts. As the
development paradigm shifted in a more neoliberal direction, Indonesian nutrition experts danced this intricate dance well, refashioning micronutrients as a
matter of "human resource management" and a "market-based" solution that
was nationally (and of course globally) appropriate. It was in this space between
globalism and developmentalism that *gizi mikro* emerged as the key term for
Indonesian policy.

While food policies were made doubly accountable to the global and the
national, they were not made accountable to Indonesian citizens. While experts
focused on hidden hunger, more obvious hunger had not disappeared. In fact,
one might say that the government's focus on micronutrient deficiencies was
misplaced in the face of the recurrence of obvious forms of hunger. News about
protein-calorie malnutrition in the eastern regions of the country in 2005, with
shocking photos of the obviously hungry, defied the "hidden" hunger framework
that was dominating Indonesian food policy. This was a wake-up call to many
food policy experts in the country (TEMPO 2005a; TEMPO 2005b; GATRA
2005). When I asked about this news, many Indonesian nutritional experts
said that they were "shocked" to see such obvious forms of malnutrition in the
country. The presence of such visible malnutrition was something that should
belong to the past. What made the condition itself so "hidden" to the experts and
the public is that it occurs among the most marginalized of the Indonesian communities, such as the children of refugees from the former East Timor (Jakarta
Post 2009; Fointuna and Maryono 2009).

As was clear at the workshop in Bogor discussed at the opening of this chapter, there was no grassroots participation in food and nutrition policymaking.
While food policy experts were busy listening to their international peers and
fine-tuning their research proposals to fit the development paradigm, they were
not questioning what had fallen outside the international development discourse or the development paradigm of the country. The Healthy and Productive Female Worker Movement in the 1990s instructed female workers to take
iron tablets to combat anemia. These women were an engine of the celebrated
Indonesian economic development. The Suharto regime tightly controlled
unionism and successfully attracted foreign direct investment by keeping an
inexpensive, docile labor force. But the working conditions were bad and subject to growing international and domestic criticism and global antisweatshop
activism. If asked about how to help their anemia, female workers probably
would have said, "Give us decent wages and pay us for overtime if you are worried about our anemia." Instead, it was iron tablets that experts decided to give

them, thus obfuscating the problematic working conditions and the development paradigm that prioritized economic growth over people's welfare.

Micronutrients came to be the focal point of the definition of the food problem in Indonesia in the 1990s. Such a shift was not propelled by demands from the hungry themselves. It was the international and national experts who saw and sought micronutrients as the best way to address food policy. The result was that efforts were focused on synchronizing Indonesian food policy with international scientific consensus and national development priorities but not with the needs of the poor and the marginalized, who were not heard.

5

BUILDING A HEALTHY INDONESIA WITH FLOUR, MSG, AND INSTANT NOODLES

Promoting the nation's nutrition (Turut membangun gizi bangsa)

—Bogasari Flour Mill slogan

Wheat flour fortified with vitamins. Built to improve the health of the nation.

—Sriboga Raturaya Flour Mill slogan

The world's largest flour mill is located in an unlikely place—in Indonesia, whose population is not known for eating bread or pasta. Bogasari Flour Mill, Indonesia's largest milling company, has the world's largest mill, located in Tanjung Priok, the industrial zone filled with warehouses and factories in the northern port of Jakarta. Its state-of-the-art mill, silos, and other equipment are decorated with the well-known Blue Key logo that is frequently seen on supermarket shelves and on billboards in town. Its capacity is enormous, far exceeding mills in the United States or Canada. Although falling far short of Bogasari's dominant presence in the market, five other flour mills in the country also have very modern milling facilities. The one I visited was in the process of installing the latest equipment from Europe.

For a country composed mainly of rice and cassava eaters, the size of Indonesia's flour milling industry is astonishing. It has the largest milling capacity in Southeast Asia, well beyond that of the Philippines, whose people one might imagine using more wheat flour. Indonesia's "traditional cuisine" is as diverse as its several hundred ethnic groups, but the people tend to use rice, cassava, and maize as their staple foods, not wheat. Indeed, Indonesia has virtually no domestic production of wheat, and most of the wheat consumed there is imported.

The story of this disproportionately large industry cannot be told without touching on Suharto's close relationships with many of the key actors in the milling industry and the economic empire that they built up during his New Order regime. For three decades, the milling industry was one of the key cash cows for Suharto and his cronies. Under their guardianship, the milling industry grew tremendously.

In the late 1990s, this powerful industry became a central actor in Indonesia's food policy. Following other developing countries that had mandatory fortification programs, the government decided on a mandatory flour fortification policy. Wheat flour became the first commodity besides iodized salt to be fortified by law in Indonesia. The new regulation required that *all* wheat flour sold in Indonesia—imported or domestic—be fortified with iron, zinc, folic acid, and B vitamins, according to the Indonesian National Standard (SNI).

This wheat flour fortification policy was considered a huge success for micronutrient advocates, who with it finally achieved a public fortification policy in the country. Indonesian nutrition experts had previously tried to establish mandatory fortification, without success. Therefore, when wheat flour fortification finally became the official policy, there was much for nutritional experts to celebrate. There was a big opening ceremony at Sriboga Raturaya Flour Mill for its first fortified product, attended by important industry members and government officials. Reflecting international interest in fortification as a development tool, foreign actors were also jubilant. UNICEF presented a letter of gratitude to the owner of the flour mill, commending him for being the first to comply with the mandatory wheat flour fortification regulation.[1] The US ambassador attended the ceremony at Bogasari Flour Mill.[2] International health experts commended Indonesia for its high awareness of the importance of micronutrients and praised it for becoming a good role model for other developing countries.

The textbook description of fortification is that it adds costs to manufacturing and therefore companies hate to see it imposed on them. The nutrition literature posits the corporate sector as the prime bottleneck for fortification that needs to be attended to by policymakers. In various countries, such literature points out, fortification attempts have been aborted due to industry opposition to the increased costs. Indonesia has had its share of this kind of experience. Other powerful industries in Indonesia successfully resisted the government's attempts to mandate the fortification of their products. Since the flour milling industry in Indonesia was not only economically powerful but also politically well connected, it is worth asking why they did not lobby against a cost-adding food policy, and so became a "victim" of a mandatory fortification program in Indonesia. Was it a heroic act of self-sacrifice for the nation's health as claimed by the industry?

In fact, the Indonesian milling industry has benefited from the wheat flour fortification program. The timing of the policy was interesting; the mandatory fortification requirement started around the time that this previously protected industry was deregulated, and imported wheat flour started to flood the Indonesian market. When the fortification regulation became law, much of the imported flour could not satisfy it, and hence could not be imported.

Of course, if you ask the people who were involved, wheat flour fortification was not meant to be an industry subsidy. It was conceived of as a public health intervention and justified as such. Yet the sheer political connectedness of the flour mill industry in Indonesia makes it natural for a casual observer to conclude that it was a classic case of science and public health objectives distorted by the economic power of the agrofood business. This also resonates with the prevalent model of science in popular writings that takes science to be a mere tool for powerful social actors. In this view, the problem is seen as the existence of a food industry that manipulates nutritional science for economic advantage.

As I traced the history of mandatory fortification and conducted interviews with stakeholders, however, it became clear that wheat flour fortification cannot be dismissed simply as a corporate takeover of science. It was not the milling industry that took the lead in fortification. Policymakers and nutritional experts had laid a good deal of ground work before those in the industry realized that fortification would benefit them as a form of trade barrier. And the experts' efforts go back more than a decade. Milling industry executives could not have cooked up the fortification plan overnight as a result of the economic crisis.

Who actually benefited is an important part of the story, but another important part—and the more interesting story here—is how it came about. Scientists and health experts worked very hard beginning in the 1980s to get fortification implemented as a public policy. For them, the political-economic implication of flour fortification—that it might aid a monopoly industry—did not seem to matter too much. Many fortification advocates are well-intentioned, smart, and dedicated scientists. When I interviewed nutritional scientists and health bureaucrats in the country on this subject, I could not help wondering why they ended up helping a powerful monopolistic industry. I wanted to explore the logic that they operated under and what led them to push for wheat flour fortification.

In an effort to make sense of the experts' support of wheat flour as healthy food, in this chapter I contextualize wheat flour as part of the longer history of the nutritional experts' network in Indonesia. I look at not only the wheat flour fortification program but also two preceding fortification efforts in Indonesia: those involving monosodium glutamate (MSG) and instant noodles. Despite dubious health properties and questions about cultural appropriateness, MSG and instant noodles came very close to becoming officially sanctioned "healthy foods." By going back to these curious pre-wheat flour cases, what stands out is not the abnormality of the wheat flour case but its continuity with previous cases.

What was the vision that translated the problem of malnourishment into a need for fortified wheat flour? For MSG? For instant noodles? How were the needs of people defined in the fortification network? By weaving together the three stories of MSG, instant noodles, and wheat flour fortification, I highlight the pervasive influence of nutritionism in defining the food "needs" and "problem" in an extremely specific manner—prioritizing quantifiability, universality, and simplicity—while simultaneously naturalizing that definition. It was this logic that was critically important in translating the problem of malnourished people into the "need" for fortified products, authorizing the public health campaigns for fortified MSG, instant noodles, and wheat flour.

Significantly, the "needs" of the malnourished and hungry were identified almost exclusively by "scientific data" and not through a democratic participatory process. It was nutritional surveys that were instrumental in pushing forward fortification projects, since experts believed that they provided undeniable evidence for the need for fortification. In contrast, throughout the three fortification attempts, little opportunity was available for ordinary citizens to discuss fortification's desirability and its social, cultural, and political implications. Women, in particular, were the presumed beneficiaries of fortified instant noodles and wheat flour, since anemia was the target. Ironically, however, ordinary women were largely absent in the two decades of fortification policy debate. This reflects the reality of how women as a biologically coded group were salient while actual women who suffered from micronutrient deficiencies were absent from the discussion.

Instead of working with poor women, nutrition and development experts worked closely with private industry. Furthermore, nutritionism's reductive focus on nutrients conferred on these corporations a status of expertise that were almost equal to that of nutritional scientists, since they were the ones who knew practical details of manufacturing, marketing, and distribution. Within the purview of nutritionism, nutritional scientists, corporate staff, and nutritional surveys were sufficient to provide necessary inputs in formulating fortification programs, while many cultural and political issues remained unaddressed.

Crony Agribusiness: The Flour Milling Industry under Suharto's New Order

The white house, which brought to mind the residence of the US president, to which I was invited by the owner of Sriboga Raturaya Flour Mill was a stark reminder that the industry was emblematic of New Order cronyism. Located near the company's mill in Semarang, the shining white mansion resembling a Greek temple stood in the middle of the residential area. The owner, Alwin Arifin, manages Sriboga Raturaya Flour Mill, one of four flour milling companies in Indonesia, but it was his father, Bustanil Arifin, who was in charge until recently. Bustanil Arifin owned two of Indonesia's milling companies at one point—the Sriboga Raturaya Flour Mill and Berdikari Flour Mill. He and his family all lived in Jakarta, and the house seemed to have no regular residents except for several servants who maintained it. No one said it was a museum, but the house had all the qualities of a museum. All the rooms were bright and clean without any dust in sight, much better kept than most of the museums in Indonesia that I had visited. In the center of the main room was a round table covered with big photos of Suharto and his wife and of Bustanil Arifin and his wife. On the walls were portraits of the ancestors of the Suhartos and the Arifins, emphasizing the royal lineages of the wives. The smaller rooms to the side also housed memorabilia of all kinds. The house's sole objective seemed to be to commemorate and remind people of the glorious lineage and achievements of the Arifins. In the age of "Reformation," when anything related to Suharto tended to be stigmatized, I found the open celebration of the link with him in this "White House" quite striking.

The White House embodies Arifin's close ties to Suharto. But he was not the only one in the milling industry with such a connection. In fact, the history of this industry in Indonesia has a familiar resemblance to other stories of Suharto's cronyism. The industry was managed by a "who's who" of Suharto's inner circle. It seems as if participation in the milling industry itself was a form of patronage, as Suharto kept adding his favorite people to it.

Until the deregulation of the industry in the 2000s, there were five mills owned by four companies in Indonesia (table 5.1); Bogasari (Jakarta and Surabaya, owned by Indofood), Sriboga Raturaya (Semarang), Berdikari Sari Utama (Ujung Pandang), and Panganmas Inti Persada (Cilacap) (Purnama 2003).

The largest and oldest mill, Bogasari, was started by Suharto's long-time friend and confidant, Liem Sioe Liong, and Suharto's cousin, Sudwikatomono (Aditjondro 2000). Liem Sioe Liong is an ethnic Chinese businessman, originally from Fujian Province in China. He successfully expanded his business to form one of Indonesia's largest conglomerates, the Salim Group, under Suharto's protection. Liem's relationship with Suharto began when he started up a trading business in Central Java. Suharto was at that time an officer in the army, stationed in

TABLE 5.1 Indonesian milling industry

COMPANY NAME	MILL LOCATION	PRODUCTION CAPACITY (METRIC TONS/DAY)	OPERATION STARTED	EMPLOYEES	ORIGINAL OWNER
Bogasari	Jakarta/Surabaya	11,766	1971	2,600	Liem Sioe Liong
Berdikari Sari Utama	Ujung Pandang	2,146	1973	484	Bustanil Arifin
Panganmas Inti Persada	Cilacap	740	1997	384	Tutut
Sriboga Raturaya	Semarang	1,100	1998	300	Bustanil Arifin

Source: Indonesian Association of Wheat Flour Producers (APTINDO) and interviews.

Central Java's Diponegoro Division. He was put in charge of supply and finance of the division, and that role connected him to the Chinese merchant Liem. Their hip grew over the years and further strengthened after Suharto rose through the ranks to become head of the military, finally replacing the Republic of Indonesia's first president, Sukarno, as acting president in the midst of the chaos caused by the alleged Communist Party coup in 1965 (Aditjondro 2000).

Suharto's New Order regime has a contradictory legacy in the management of the economy, and Liem is a quintessential example of its darker side. On the one hand, immediately after the coup, Suharto successfully controlled the wild inflation of the time and restored international business confidence, which was faltering partly due to Sukarno's nationalistic programs and anti-West rhetoric. When the clearly anti-Communist Suharto took over, Indonesia enjoyed an influx of Western aid and foreign direct investment. Oil and natural gas revenues also helped the national economy, and the GNP grew by about 4–5% per annum. The other side of the prosperous New Order economy, however, was the increasing takeover by Suharto's business allies and his own family. Suharto created a system of favoritism and cronyism among a handful of ethnic Chinese businessmen and his own family members. Some Chinese-owned businesses expanded tremendously under Suharto's protection and favor, growing into the country's major conglomerates such as the Astra, Sinar Mas, and Lippo groups.

Liem's Salim Group was one of those conglomerates. Liem helped Suharto in the early years of his presidency by investing in infrastructure projects when the cash-short government could not. In return, Liem received special incentives and financial deals (Schwarz and Friedland 1991). His business, the Salim Group, grew tremendously and eventually came to hold the country's largest market share in various key sectors, including processed foods, private banking, cement, several grain commodities, auto manufacturing, chemicals, and real estate. By 1990, the Salim Group had revenues of $8–9 billion, with its domestic sales equivalent to 5 percent of Indonesia's GDP. In the 1990s, it controlled three hundred companies employing 135,000 Indonesians (Friend 2003). The Salim Group expanded overseas as well, buying businesses in Singapore, Hong Kong, the Philippines, and Australia, and elsewhere (Schwarz and Friedland 1991).

Suharto himself was intertwined with the Salim Group's business fortunes. In return for his political patronage, Suharto received a percentage of the profits. According to George Aditjondro (1998), there were four investors who divided Salim Group's profits. Suharto's foster brother, Sudwikatmono, was one of them, along with Liem himself; Djuhar Sutanto, a Chinese businessman; and Ibrahim Risjad, an Achenese with close ties to the military. They usually divided the profits of investment with 40 percent each for Liem and Djuhar, and 10 percent each for Sudwikatmono and Risjad. Aside from personal ties, this arrangement ensured that Suharto and the Salim Group's relationship was close and tight.

It is indicative of these close relations that Suharto himself came to the opening ceremony, cut the ribbon, and celebrated Indonesia's first modern mill when Liem opened Bogasari Flour Mill in Jakarta in May 1971. The picture of that event still decorates Bogasari's corporate brochure.

Flour milling was a cash cow for the Salim Group. Bogasari Flour Mill obtained a right of monopoly on wheat imports and flour milling from Suharto. The profitability of Bogasari was also rooted in the fact that the wheat was supplied under the United States' foreign aid program Public Law 480 (the Food for Peace program) on concessionary terms (Aditjondro 2000). PL 480 was a food aid program for developing countries that was started in 1954, its purpose was both to aid in development and to dispose of US agricultural surpluses and create new overseas markets for US agriculture. As in many other cases, wheat for Indonesia under PL 480 was part of international Cold War politics. The United States had been irritated by the anti-West tendencies of Suharto's predecessor and had witnessed with concern his growing relationship with the Communist sphere. The United States therefore welcomed the transition from Sukarno's Guided Democracy to Suharto's New Order. The US government, along with that of its Western allies, rewarded Suharto handsomely with huge amounts of foreign aid; the wheat donation was one of these rewards. Under PL 480, wheat was provided through long-term loans with reduced interest rates. Suharto even received some special exemptions from regular PL 480 requirements (Magiera 1993).

With cheap raw materials, government subsidies, and political backing, Bogasari expanded greatly. It added a second flour mill in Surabaya in 1972. Bogasari

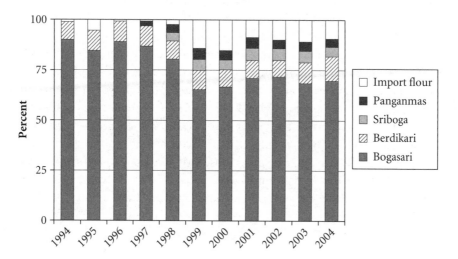

FIGURE 5.1. Market share trend of wheat flour market in Indonesia.
Source: APTINDO.

has since maintained the largest market share of wheat flour in Indonesia, capturing more than 70 percent of it (fig. 5.1).

Indonesia's second milling company, Berdikari, was managed by Bustanil Arifin, who is the owner of the glorious White House. Berdikari's history goes back to 1970, when a company from Singapore (Prima Limited) opened a mill in Makassar, in South Sulawesi. It was initially managed by Bogasari. In 1982, the founding shareholders sold their entire interest to a state-owned company called PT Perusahaan Pilot Projek Berdikari (PT PP Berdikari), headed by Bustanil Arifin. Arifin later managed to turn this state company into a private firm of his own (Tiwon 1999). Bustanil Arifin was another of Suharto's confidants. He was an Achenese military general, and he was tied to Suharto by family—Arifin was married to a relative of Suharto's wife, Christine. He became a close partner of Suharto, centrally involved in most of his money-making schemes, and has been called the "single most important fund-raiser" for the Suharto regime (Schwarz and Friedland 1991). Not only did he control Berdikari beginning in the 1980s, but he later was put in charge of the Food Logistics Agency (BULOG), a powerful state agency originally set up to stabilize the price of rice and to distribute it, but which gradually expanded to control other key commodities including wheat and sugar. Arifin was also appointed Minister of Cooperatives, which was one of the money-making machines of New Order cronyism. In addition, he sat on the boards of Suharto's foundations (*yayasan*), such as the Indonesian Institute of Management Development and the Indonesian Institute of Cooperatives (Tiwon 1999). These foundations were also well known as tools that Suharto and his cronies used to make money through corrupt activities. For instance, under Arifin's management, BULOG channeled large sums from state accounts to these Suharto-related foundations (Jakarta Post 2000).

Arifin controlled Berdikari Flour Mill, first through PT PP Berdikari, and later by becoming the chairman of the Berdikari mill (Tiwon 1999). The mill was renamed Berdikari Sari Utama Flour Mills in 1983. Perhaps Arifin saw that the milling industry was so profitable that he had to have another company for himself. In 1998, he opened Sriboga Raturaya Flour Mill on his own in Central Java. This is the third-largest milling company in Indonesia and now is under the control of Arifin's son.

The fourth flour mill, Panganmas, was started in 1997 by Tutut, Suharto's eldest daughter. Like other members of the Suharto family, Tutut amassed huge wealth under a corporate group called Citra Lamtoro Gung Group. This conglomerate controlled a range of economic activities, including a toll-road company, telecommunications, banking, plantations, construction, forestry, sugar refining, and trading (Comey and Liebhold 1999). In 1997, Citra received a permit from the government to open up a new flour mill.[3]

In sum, what might at first glance seem like a healthy industry with four competitors was in reality made up of a small group of Suharto's favorites. If one digs a little further, one discovers that the cross-ownership of shares was so intertwined that the industry was essentially one gigantic monopolistic corporation. According to Aditjondro (1998), PT PP Berdikari was owned by Bustanil Arifin (30%), Salim (40%), and Bob Hassan (30%). Bob Hassan was another infamous crony of Suharto's, who was mainly implicated in the corrupt timber business. His business empire involved not only timber but also mining, manufacturing, and social charitable organizations, much like the foundations that Arifin was involved with. Sriboga Raturaya Flour Mill's ownership was split between Arifin (75%) and Salim (25%). Arifin's wife was a majority shareholder of Bogasari from 1977 (Aditjondro 1998).

Under Suharto's protection, the wheat flour business was strictly regulated. There was an elaborate system to ensure each flour mill's prosperity. After 1972, BULOG controlled all aspects of the wheat flour business. As Indonesia does not produce wheat, all wheat was imported, and BULOG was given control over all wheat imports. Technically, it was BULOG that imported wheat, and each flour mill just milled grains for BULOG and received fees for milling it (USDA 1997). BULOG also kept tight control on the marketing of wheat flour. All flour distributors had to be approved by BULOG, organized under the Association of Sugar and Flour Distributors, and they had fixed territories, except for some bread and noodles cooperatives, which were allowed to obtain wheat flour directly from BULOG (Fabiosa 2006).

What this system meant for the milling companies was guaranteed profitability. Unlike mills in other countries, Indonesian mills did not have the risk of trading grains by themselves. In addition, because the government fixed the milling fee at a generous rate, the profit margin for Indonesian mills was high. For instance, economist Stephen Magiera (1993) calculated that in 1988, the total mill margin was $35.68 per ton in Indonesia, compared to the $10 margin for typical mills in the United States. Besides this lucrative "milling fee," mills had another source of revenue: they were allowed to keep all milling by-products, which could be sold as animal feed (Fabiosa 2006). The mills received revenues of about $38 per ton from the sale of milling by-products (Magiera 1993).

In addition to the four flour mills, other food companies also benefited from the patronage system. Most notable was Indofood, which was also owned by the Salim Group.[4] Indofood is one of the largest food manufacturers in Asia, with 45,000 employees, and is best known for its instant noodles, of which it sells about nine billion packs every year. Indofood had a great advantage because the company could procure wheat flour from its sister company, Bogasari. Their

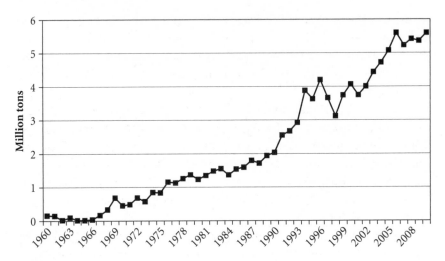

FIGURE 5.2. Indonesian wheat imports, 1960–2010.

buying price for wheat flour was much lower than the global market price (Magiera 1993); their competitors had to pay distributors who paid a "surcharge" to BULOG. It is perhaps thanks to the cheaper raw material that Indofood became Indonesia's largest food processor and, in fact, the world's largest instant noodles producer. Indofood's market presence is overwhelming; it has more than a 90 percent market share of instant noodles in Indonesia, and uses a significant portion of the wheat imported by the country (Purnama 2000, 71). In sum, it could be argued that the biggest beneficiary of this scheme was the Salim Group, which owned both Bogasari Flour Mill and Indofood (Kwok 1997; Gozal 1998).

Under the New Order's protection, Indonesia's wheat imports grew from less than a half a million tons in 1974 to more than four million tons in the mid-1990s (fig. 5.2). All of the wheat was milled by these four companies.[5] The country's total wheat flour milling capacity grew tremendously and is now much bigger than capacities in other Southeast Asian countries (tables 5.2 and 5.3).[6]

Fortification Policy: A Lifesaver for a Monopoly Industry

The glorious days of Suharto's New Order came to a halt when the Asian Financial Crisis hit first in Thailand and then spread across the region. Indonesian currency fell from Rp 2,400 per dollar in June 1997 to Rp 16,000 per dollar in June 1998. Economic upheaval sparked a political crisis. Discontent with the New

TABLE 5.2 Flour production capacity in Southeast Asian countries

COUNTRY	FLOUR PRODUCTION CAPACITY (METRIC TONS/YEAR)	MILLS
Indonesia	4,728,600	Bogasari, Berdikari, Sriboga, Panganmas
Philippines	2,673,620	General Milling, Philippine Foremost, Wellington, Morning Star, Purefoods, Pilmico, Republic, Universal Robina, Liberty, Delta Milling Industries, Philippines, Pacific, Nissin Monde
Malaysia	1,269,840	Federal, Malayan, Kuantan, Seberang, United Malaysian, Lahad Datu, Sarawak, Sabah Flour & Feed
Thailand	925,000	United, Laemthong, Siam, CP Thai, Bangkok, Kerry/Thai President, Nisshin-STC
Singapore	199,800	Prima Flour

Source: APTINDO.

TABLE 5.3 Top ten flour mills in the world by capacity

RANK	MILL	COUNTRY	CAPACITY (METRIC TONS/DAY)
1	Bogasari Flour Mills–Jakarta	Indonesia	7,400
2	Bogasari Flour Mills–Surabaya	Indonesia	4,366
3	Prima Flour Mills	Sri Lanka	2,600
4	Berdikari Sari Utama	Indonesia	2,146
5	Nabisco Brands	USA	1,600
6	ConAgra Flour Milling	USA	1,450
7	General Mills	USA	1,300
8	ADM Milling Corp.	Canada	1,200
9	Sriboga Raturaya Flour Mills	Indonesia	1,100
10	General Milling Corp.	Philippines	1,100

Source: APTINDO.

Order's cronyism had already been close to erupting. Students took to the streets demanding economic and political reforms. Unable to cope with the economic shock and increasingly threatened by the political instability, Suharto asked for IMF bailouts in October 1997 and January 1998. The IMF offered to provide new credit in return for major economic reforms (Liddle 1999). Initially resentful of these economic reforms that would destroy his and his cronies' economic empires, Suharto grudgingly had to agree to the IMF terms. One of the IMF's terms is of particular importance here: the requirement to open up the flour industry. As a part of the package, the IMF required BULOG to release control over the wheat-import business. The IMF required the government to eliminate tariffs on wheat imports along with other commodities and to allow free competition in importation of wheat and wheat flour and sale or distribution of flour (IMF 1998; USDA 1997).

These reforms meant a sea change for the hitherto protected industry. Importation of wheat was deregulated, and BULOG was no longer the only importer. Now, independent traders and food processors could import wheat flour directly. The import duty on wheat flour was also reduced to zero in 2000. As a result of these economic reform measures, the Indonesian milling industry faced tough competition for the first time in its history. The biggest threat to the Indonesian milling industry was imported wheat flour. As the subsidy declined and the duty on wheat flour dropped, the mills faced growing competition from foreign flours. The impact was acute. Only six months after the market was liberalized, Indonesia imported a substantial 150,000 metric tons of flour (Government of Australia 2000). Wheat flour imports had been at a level of 22,000 tons per year before the reform, but this increased to 500,000 tons after the liberalization (APTINDO 2001). Of course, the industry did not remain beaten. The milling industry resorted to various means to curb the impact. They argued that the import companies were dumping in Indonesia and asked the Indonesian Anti-Dumping Commission to conduct an investigation into the dumping of wheat flour onto the Indonesian market beginning in 2000. The industry also lobbied the government to raise the import duty again.[7]

As it turns out, Indonesian nutrition experts and the flour industry had been working on the fortification policy for a while. In a climate of competition unprecedented in the history of the Indonesian milling industry, companies found a benefit in fortification: a trade barrier that was justifiable on public health grounds. The fortification policy made it more difficult to import wheat flour, because the Indonesian National Standard for wheat flour fortification is different than any other country's (table 5.4). The wheat flour milling industry therefore found that they could be protected from imported wheat flour by having a fortification requirement. This is not a secret. In many interviews I had with nutritional experts, they suggested this was the key ingredient for the successful fortification policy. For instance, a former official in the Office of State Minister of Food Affairs said, "Making it mandatory was also a need of industry itself. Why? Because with it, the quality of imported wheat flour from Australia or elsewhere becomes not good, or lower because they are not fortified. Therefore the industry feels, with this [fortification], there is a certain policy instrument for making some sort of technical barrier."[8] The head of the nutrition division in the Ministry of Health also recalled that the industry was interested in fortification "because they were worried about globalization. They said if they fortify the wheat flour, then another wheat flour cannot enter in Indonesia. That's what they were worried about. We got pressure from foreign competitors."[9]

TABLE 5.4 Wheat flour mandatory fortification standards (in ppm)

COUNTRY	VITAMIN B$_1$	VITAMIN B$_2$	FOLIC ACID	NIACIN	ZINC	IRON
Bahrain			1.5			
Belize	4	2.5	1.5	45		60
Bolivia	4.45	2.65	1.5	35.6		60
Canada	6.4	4.0	1.5	53		44
Chile	6.3	1.3	2.2	13		30
Columbia	6.0	4.0	1.54	55		44
Costa Rica	6.2	4.2	1.8	55		55
Cuba	7.0	7.0	2.5	70		45
Ecuador	4.0	7.0	0.6	40		55
El Salvador	6.2	4.2	1.8	55		55
Guatemala	6.2	4.2	1.8	55		55
Honduras	6.2	4.2	1.8	55		55
Indonesia	2.5	4	2		30	50
Jordan			1.5			30
Kuwait	6.38	3.96	1.5	53		44
Mexico	4.0	2.4	1.6	28	16	24
Nicaragua	6.2	4.2	1.8	55		55
Nigeria	6.2	3.7		49.5		40.7
Oman			1.5			30
Panama	6.0	4.0	1.5	55		60
Paraguay	4.5	2.5	3.0	35		45
Peru						28
Qatar			1.5			60
Saudi Arabia	6.38	3.96	1.5	52.9		36.3
South Africa	1.94	1.78	1.43	23.7	15	35
Trinidad Tobago			1.5			30
UAE			1.5			30
UK	2.4				16	16.5

Source: www.sph.emory.edu/wheatflour/training(resources/fortstds2.pdf), data as of 2002.

The milling industry started to give serious thought to a fortification policy. It increasingly seemed like a good idea. Seeing the industry more willing than in this area, nutritional experts were jubilant. Details were worked out, and the official law for mandatory fortification of flour was issued in 2001. The effect was dramatic. In 2001, imports of flour decreased radically (fig. 5.3). The Indonesian milling industry wanted to make sure that the regulation would continue to block their foreign competitors. Even after the initiation of the fortification law, they repeatedly complained that most imported flour was still not fortified in compliance with the SNI and urged the government to conduct stricter monitoring. The Indonesian Association of Wheat Flour Producers' (APTINDO) spokesperson, Ratna Sari Loppies, frequently appeared in the media, claiming that much imported wheat flour did not satisfy SNI or labeling requirements and that the

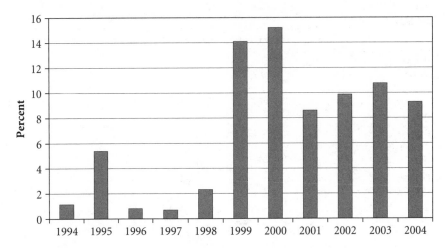

FIGURE 5.3.   Market share of imported wheat flour in Indonesia.
Source: APTINDO.

domestic industry was subject to unfair competition (see, e.g., Kompas 2003a, 2003b, 2003c). Their lobbying was partially successful. Several years after the start of the policy, the government issued a new regulation requiring imported flour to be registered with the Ministry of Health.[10]

There were some criticisms of this policy. Even within the government, some considered this a thinly veiled protectionist policy posing as an antimalnutrition policy. Particularly vocal opposition came from the Business Competition Supervisory Commission (Komisi Pengawas Persaingan Usaha or KPPU). Its head, Soetrisno Iwantono, criticized mandatory fortification of wheat flour, saying that it might constitute an unfair trade barrier. Another member of KPPU, Ani Pudyastuti, said that mandatory fortification in SNI was an entry barrier, and pointed out that there were no clear benefits from the added micronutrients (Kompas 2003b).

The wheat flour industry fought back. APTINDO dismissed the KPPU's charge of creating an unfair trade barrier and said that it was not the job of the KPPU to investigate fortification policy (Kompas 2003b). Furthermore, the milling industry launched a public relations campaign, praising the fortification program as a nutritional endeavor and portraying itself as sacrificing for the national development. For instance Sriboga Raturaya Flour Mill distributed promotional material headed "Wheat Flour fortified with vitamins. Built to improve the health of the nation" (PT Sriboga Raturaya n.d.). Similarly, the industry organization, APTINDO, argued:

Even today, many imported wheat flour products still come to the domestic market that do not fulfill the Indonesian National Standard requirement, although the government already has decided that the wheat standard is mandatory and that it needs to be fulfilled by every company that wants to market its products in Indonesia. This situation already has caused losses to the national wheat flour industry as it has implemented the fortification program that constitutes one of the decisions of SNI. Domestic producers need to pay an additional R40 million per year. (Ministry of Trade and Industry 2003, 49; my translation)

The industry emphasized the cost it was bearing for fortification and the high moral position it was taking in the name of the nation's health, and pleaded for a crackdown on the foreign competitors.

However, in reality, the fortification required very little new investment and the micronutrient premix's price was modest. Interviews with industry insiders confirmed that cost was actually not a major problem for the industry. I asked managers at the mills what was entailed in adding extra nutrients, and they said that changes in processing were not consequential at all. Yet in their self-portrayal, the Indonesian milling industry was making a heroic sacrifice for the nutrition of the nation and was the victim of foreigners who were dumping cheaper wheat flour on the market.

MSG and Instant Noodles Are Good for You

This analysis of the milling industry raises many questions as to the actual motivation for the fortification regulation. The description of the industry and the impact of wheat flour fortification could imply that fortification was implemented to benefit the powerful monopolistic industry. Yet this conclusion, albeit attractive due to its straightforwardness, would undermine a realistic assessment of nutritional science on the ground. For one thing, exploration of flour fortification had already begun before the economic crisis and liberalization. The country's nutrition experts had lobbied for wheat flour fortification well before it became financially valuable for the industry. Therefore, it would be inaccurate to assert that wheat flour fortification was only driven by corporate interests. We also need to understand that the technoscientific community's support for fortification could not be forged overnight at the will of the corporations, however powerful they were. Over the decades leading up to 1998, the Indonesian nutrition experts had built up a network of institutions and scientists interested in fortification. Particularly strong proponents emerged in the national nutrition research institutes that conducted multiple studies on possible vehicles beginning in the early 1980s. Many of these experts had worked with international vitamin A researchers such as Alfred Sommer. Although they initially strongly supported the supplement approach, namely vitamin A pills, these experts expanded their interests to other tools to address the problem. The network enjoyed strong support from international organizations and, over the years, gained sympathizers within the Ministry of Health as well. We need a better framework than the simple contamination model of science to understand how these technoscientific experts ended up helping the crony industry.

Wheat flour's curious predecessors in fortification drives were instant noodles and monosodium glutamate (MSG) (see table 5.5). MSG was one of the

TABLE 5.5   Chronology of fortification projects in Indonesia

1985	Research on MSG fortification (Muhilal and Murdiana 1985)
1988	Research on efficacy of fortified MSG (Muhilal et al. 1988)
1989	Five-Year Development Plan (Repelita) III mentions fortification
1991	Research on instant noodle fortification (Soetrisno, Slamet, and Hermana 1991)
1993	Research on wheat flour and condiment fortification's impact on the final products' characteristics (Komari and Heraman 1993)
	National Workshop on Food and Nutrition (Widyakarya Nasional Pangan dan Gizi) includes one panel on fortification
1994	Fortified rice project started
1995	National Household Health Survey (SKRT)
	Research on instant noodle fortification with vitamin A and iron for children under five (Sukati et al. 1995)
	Research on instant noodle fortification with vitamin A and iron for pregnant women (Saidin et al. 1995)
1996	Office of State Minister of Food Affairs established. Food Law mentions the needs of fortification.
1997	National Fortification Committee established
	Research on standard fortification level for iron-fortified wheat flour (Muhilal, Murdiana, and Hermana 1997)
1998	Widyakarya discusses fortification
	With funding from UNICEF, USAID, and CIDA, testing of wheat flour fortification in a factory.
	Ministry of Health decree on wheat flour fortification (32/MENKES/SK/VI/1998)
2001	Ministry of Industry and Trade issues the Indonesian National Standard (SNI) for wheat flour
	Indonesian government notifies WTO about health-based trade restrictions, which are accepted

earliest serious fortification attempts by nutrition experts and the government in Indonesia, beginning in the 1980s. In the 1990s, researchers contemplated instant noodles as another possibility.[11] And these were not isolated laboratory experiments by overly enthusiastic scientists. They actually came very close to making it into national policy. These two cases offer an interesting opportunity to understand the dynamics of fortification. In particular, the MSG and instant noodles cases are instructive for two reasons. First, with their similarity and continuity of logic with wheat flour fortification, they demonstrate that wheat flour fortification should not be considered a historical accident. The narrow view of the "needs" of the people in these cases, I suggest, is symptomatic of the way that food problems are defined within nutritionism. These two cases exemplify the reductionist logic that circumscribed the description and prescription of the malnutrition issue, with striking resemblances to the case of wheat flour fortification. Second, they show that fortification is not only driven by the logic of science, nor only by the logic of the market, but by a coalescence of both. Only when industry was successfully enrolled in the endeavor was fortification possible. MSG and instant noodles could not bring the industry into the network, while the convergence of two logics led to the success of wheat flour fortification. This leads to the question: How did industry, not citizens, become the veto power in fortification? Ultimately, MSG, instant noodles, and wheat flour fortification were viewed as unproblematic due to various characteristics of nutritionism in which the problem is interpreted as one of nutrients rather than as a social problem; where statistics are presented as the only valid means of knowing the problem; and where industry and technoscientific experts are considered as the exclusive authorities in defining the problem and offering the solution.

The Case of MSG

The most revealing story of fortification in Indonesia before wheat flour is that of MSG. In the earliest fortification attempt by nutrition experts in the country, they wanted to add vitamin A to MSG in order to reduce vitamin A deficiency. The initial impetus for the MSG program came from a survey that calculated the prevalence of vitamin A deficiency in the country. In order to tackle the problem, government and international organizations initially relied on the mass distribution of vitamin A supplements, but they also started to look at other options, including fortification. The government hired consultants, including Alfred Sommer, whose studies in Indonesia on vitamin A deficiency had had a major international impact (Edmunds 1989). On the Indonesian side, Muhilal at the national laboratory became the foremost researcher on fortification. In

1980, the Ministry of Health began studies to identify an appropriate "vehicle," that is, a food product to be fortified for the purpose of vitamin A deficiency eradication. They funded a household consumption survey to quantify national dietary patterns and to identify candidates for a vitamin A fortification vehicle. From this survey, researchers found several food items with potential: salt, MSG, wheat, and sugar. Soon after, Muhilal and his collaborators started a pilot project on fortifying MSG (Muhilal and Murdiana 1985).

These researchers at the national nutrition research institutes found that forti-fication of MSG was technically feasible and relatively easy. Adding vitamin A did not disturb the product's characteristics too much, and it remained stable during storage and was amenable to mass production. After sorting out the technical side of fortification, researchers then went on to demonstrate that it worked by conducting a small nutrition efficacy study. To their delight, the study showed that fortified MSG improved the nutritional indicators. When consumed, the fortified MSG increased the level of serum retinol in the blood. Researchers did not forget to underline in their report that, in particular, young children and pregnant women benefited from this fortification (Muhilal et al. 1988). The forti-fied MSG project seemed like a great success, and the researchers who had con-ducted the study, particularly Muhilal, became ardent supporters.

When I interviewed them, the MSG researchers did not have any regrets about having chosen MSG for fortification. The choice was, in their view, justified by the data—they had survey data on the prevalence of the deficiency and on food consumption patterns. There was a deficiency, and people consumed MSG. The only remaining issue was to find out whether it was technically feasible and whether it would satisfy the nutrition efficacy study. Outside of the framework of the project, however, the choice of MSG might have seemed strange for two reasons. First, MSG is not a traditional part of the Indonesian diet. Glutamic acid was discovered by a Japanese researcher as a flavor enhancer at the beginning of the twentieth century, and subsequently MSG was mass produced by a Japanese food conglomerate, Ajinomoto. Ajinomoto aggressively marketed MSG not only in Japan but also in other Asian countries, successfully making it ubiquitous in Asian kitchens (Consumers Association of Penang 1986; Dibb 1999). Govern-ment fortification would mean that this multinational business product would be promoted as a healthy ingredient and that regular consumption would be encouraged throughout the country. The cultural and social implications of such a policy were not debated.

Additionally, given that there had been controversy around MSG's negative health effects, it was indeed remarkable that MSG was chosen to become the national healthy food. Chinese restaurant syndrome had already been reported, causing much controversy within and beyond medical circles worldwide. Researchers reported MSG-related symptoms such as flushing, tightness in the chest, increased blood pressure (Kwok 1968), and headache (Kenney and Tidball 1972; Schaumburg et al. 1968). Consumer concern was so great that in the late 1970s, manufacturers "voluntarily" removed MSG from baby food. Despite the controversy, nutritional researchers insisted that fortified MSG could benefit "vulnerable" groups, namely, children, pregnant women, and lactating women.

It might be argued that current research tells us that adverse health effects of MSG have turned out to be negligible. For instance, the 1988 Joint FAO/WHO Expert Committee on Food Additives decided that there was no significant health threat from MSG, and a similar conclusion was reached by the European Union and the American Food and Drug Administration. However, the FDA still admits that there are particularly sensitive subgroups within the population (Walker and Lupien 2000). In addition, these reassuring statements did not come until the late 1980s, thereby making the MSG espousal in Indonesia in the mid-1980s unusual. Yet these health concerns with MSG were largely carved away because of the reductionist framing of the issue.

It was not the realization that an otherwise "unhealthy" food could not be made "healthy" by fortification that stopped the MSG project from becoming a reality. Nor was it the realization that this might not be a real solution to the complex issue of malnutrition. It was the opposition of the MSG industry, which disliked the fortified MSG's change in color. When fortified MSG was hung in its small cellophane packets outside small rural shops—the usual way that this product was marketed and probably the reason for its successful penetration of the market—discoloration occurred. The resultant yellowish color was unacceptable for the producers, who had marketed MSG as the whitest of white products. Although the industry initially had celebrated the prospect of marketing their product as "healthy food" and "new and improved," it eventually backed out of the program (Darnton-Hill and Nalubola 2002).

The Case of Instant Noodles

Despite the failure of the MSG project, nutrition experts continued to eye fortification as a public policy option. Vitamin A deficiency seemed to be controlled well by pills, so the next target became iron. In the 1990s, Indonesian researchers started several experiments with iron fortification. The issue was which food item should be chosen as a vehicle for iron, and it seems that researchers soon decided that instant noodles was the best way to deliver iron. Already in 1991, researchers had experimented with iron fortification of instant noodles (Soetrisno, Slamet, and Hermana 1991). The experiments were successful, with no technical glitches. Researchers proceeded to conduct a study in South Kalimantan and South Sulawesi demonstrating that instant noodles were consumed in nearly all households in both areas. The bonus finding was that the poor consumed even more than the average consumer (Melse-Boonstra et al. 2000, cited in Darnton-Hill and Nalubola 2002). After establishing instant noodles as the ideal vehicle, it was time for a technical feasibility and efficacy study. In 1994, experts conducted an experiment by giving instant noodles fortified with iron to pregnant mothers (Saidin et al. 1995) and children under five years old (Sukati et al. 1995). These studies found fortified noodles were sufficiently effective.

From the experts' point of view, these experiments and surveys provided enough justification for making fortification of instant noodles a public health policy. Nutritional experts started to lobby government officials to fund the project. Once we step outside the worldview of experts, however, instant noodles also seem like an odd choice for a public health policy. Instant noodles are quintessential modern junk food, without significant nutrients and with many unhealthy ingredients. It is not only high-minded Western consumers who are worried about instant noodles. Already in the 1990s, Indonesian consumers were concerned about the noodles' nutritional quality, their high sodium content, and the use of preservatives and additives, among other issues. Social critics were worried about the cultural implications of the rapid increased use of instant noodles in Indonesian social and cultural life (Eviandaru 2001).

In contrast, the rise of instant noodles was good news for nutritional scientists who were working on fortification. More instant noodles consumed meant more iron delivered. A Ministry of Health official recalled the project:

> In 1994, we tried to conduct a small study first to try to involve one of the producers of instant noodles. For the first time, we tried instant noodles. Because we had data that showed consumption of instant noodles among people in rural areas was actually increasing. At the time, [there was] something like a boom of instant noodles production in Indonesia. And instant noodles were very, very cheap. They are easy to prepare. Even for breakfast, lunch, for school children, they can prepare it very easily. So when we were first thinking about how to fortify, it was instant noodles.[12]

For companies that make instant noodles, fortification was appealing because they could market their products as healthy food. The industry itself had tried to convince consumers of the products' safety and quality for some time, and they expected that a public fortification policy would boost the legitimacy of their health claims. The same interviewee from the Ministry of Health said:

> The Ministry of Health issued a ministerial decree on noodles fortification, but at the time, it was not yet mandatory. But we were just asking the producers to fortify for the health of the people. And it works,

and the producers actually also tried to promote their instant noodles. I mean the value of the product increased because of fortifying. So they always put [about nutritional benefits] in advertising.[13]

Despite the endorsement from experts, the instant noodle fortification project did not materialize in the end. The industry finally decided to oppose it, citing the increase in cost entailed by fortification. They wrote to the government that it would be an unbearable burden for them. The Ministry of Health tried to convince them that only several rupiah would be added to the cost, but still the industry was not happy with the increase.[14] Once the industry opted out, researchers had to give up the dream of healthy instant noodles.

What do these two cases of MSG and instant noodles fortifications tell us? The intriguing aspect of these attempts was how the scientific quest for nutritious food somehow ended up with a technical fix that had dubious social and cultural implications. In this regard there is a striking resemblance to the project of wheat flour fortification. Notice how "nutritional needs" were interpreted in the technoscience network. Although the overall goal was to improve the nutritional status of the population, the needs were reduced to one nutrient at a time— vitamin A or iron. The focus on a single nutrient effectively put a boundary on subsequent efforts in terms of scope and range of options. Correspondingly, the solution was a simple one, of just adding the missing nutrient to a food vehicle. The research task involved two simple steps: looking for a carrier for a nutrient and adding the nutrient to it. In this logic, MSG and instant noodles stood out as ideal products to improve nutritional status.

For experts, fortified MSG and instant noodles were brilliant solutions that encountered unfortunate technical glitches and cost problems. This sentiment is particularly evident among nutrition experts who were directly involved in the experiments. They tend to emphasize that fortified MSG and instant noodles were effective in trials and would have been a great policy if there had not been industry opposition (see, e.g., Edmunds 1989).

Although it is questionable whether MSG and instant noodles deserved official promotion as healthy food, nutritional experts did not believe it was necessary to address the cultural, political, and social implications of their work. They were following a typical protocol in which the complex reality of malnutrition was categorized into a set of data: the problem for the malnourished was identified and specified by a nutrition survey; the adequacy of the solution was confirmed by an efficacy trial. From this viewpoint, it is no wonder that little debate took place on the broader merits of MSG or instant noodles among experts. In their view, MSG and noodles were mere "carriers" of the nutrient, and what was important was that they carried nutrients to people. Nothing else. We can now understand why experts took the increasing consumption of MSG and instant noodles across socioeconomic strata as a trend to be welcomed. Making nutritional composition the only issue that ultimately mattered, nutritionism narrowed the food policy discourse in such a way as to block out broader and more complicated issues.

Expanding the Fortification Network

It was after these two failed fortification attempts that wheat flour fortification emerged and materialized. The earlier part of this chapter focused only on the milling industry. Now we know that there already had been a network of scientists and policymakers involved in earlier fortification attempts. This points to the need for broadening the examination of wheat flour fortification beyond the corporate world. Here we look in more detail at how the technoscience network for wheat flour fortification expanded, leading to the adoption of mandatory fortification policy by the government in the late 1990s.

Even after the two aborted attempts at fortification, the experts' network did not disappear. On the contrary, there was an increasing interest in micronutrients globally. As I have described, "micronutrient deficiencies" started to appear frequently in the vocabulary of development and food policy specialists, and fortification itself gained stronger momentum in the international and domestic scenes. The international development community saw Indonesia as a place fit for fortification initiatives and started exploratory projects. For instance, the Micronutrient Initiative chose Indonesia to experiment with rice fortification in the early 1990s. With MI funding, an organization called the Program for Appropriate Technology for Health (PATH) implemented a feasibility study of vitamin A–fortified rice called Ultra Rice between 1994 and 1996 (PATH 2000). UNICEF also was a source of international encouragement for fortification. In Indonesia, UNICEF had maintained good connections and working relationships with the government on health and nutrition issues over several decades. As international interest in fortification grew, UNICEF started to put fortification on its Indonesian office's agenda as well, seeking to heighten the interest among various sectors of Indonesian society. It sponsored workshops such as "A Dialogue on Food Fortification," which was held in 1996 at BULOG (Direktorat Bina Gizi Masyarakat 1997). Such workshops demonstrated the growing international interest in fortification and encouraged Indonesians to seriously consider it for public policy.

In addition to the international sponsors, a bureaucratic sponsor was also critical in the process. The emergent fortification network got a boost when the Office of State Minister of Food Affairs was established in 1996. The new ministry was created to fill the perceived institutional gap that existed between the Ministry of Agriculture and the Ministry of Health. The newly established ministry took on fortification as a mission well suited to its mandate. Fortification seemed to fall nicely within the new ministry's jurisdiction of "food": something that was neither agriculture nor health. In particular, many of my interviewees identified the assistant to the State Minister of Food Affairs for the Division of Food, Suroso Natakusuma, as the prime mover behind fortification.

This new ministry's first important job was to draft food-related laws. The resulting Food Law of 1996 was the government's first major codification on issues of food quality, food safety, and labeling, and fortification was included as an important policy option to be considered. In order to materialize its commitment to fortification, the ministry then established the Food Fortification Committee (Komisi Fortifikasi Pangan).[15] The committee invited experts, government agencies, food producers, and other stakeholders to chart the way for a national fortification policy.

In the 1990s, nutritional experts also sought more data on micronutrients and micronutrient deficiencies. Critical in creating the momentum that eventually led to wheat flour fortification was the national nutrition survey data on anemia. In particular, nutritional experts used the National Household Health Survey on anemia among pregnant women and children under five years old to demonstrate the seriousness of anemia in the country and to boost the fortification movement.[16] The survey helped the fortification network claim "scientific evidence" regarding the need for additional iron and to move on to a discussion of which vehicle to use. By this time, instant noodles had been vetoed as a candidate for fortification, and the experts had to look for other options. Rice, sugar, and cooking oil had potential, and many other countries had already used them. But each had problems in the Indonesian context—rice mills were too numerous to ensure quality control and also were deemed "too political"; cooking oil was technically difficult, as Indonesian cuisine frequently uses deep frying, which destroys some nutrients (Untoro 2002).[17] The production of sugar was considered too dispersed for fortification control and monitoring.

In contrast, wheat flour seemed to fulfill necessary conditions. The official justifications for choosing wheat flour as a fortification vehicle encompass a variety of issues, but typically they include the following points: (1) production is centralized (Natakusuma 1998; UNICEF 2003); (2) it is consumed by many people, and particularly by the poor (Natakusuma 1998); (3) its distribution is widespread, reaching remote areas (Natakusuma 1998); (4) it is affordable (Natakusuma 1998); (5) its fortification is technologically feasible (Natakusuma 1998; UNICEF 2003); (6) it would mean that instant noodles would also be fortified, and their consumption is widespread even in rural areas and among the urban poor (three times a week, and covering 80% of children two years of age) (Soekirman et al. 2005); (7) fortification adds an insignificant cost (Soekirman et al. 2005; UNICEF 2003; Soekirman 1998).

One expert echoed the official explanations when I asked why wheat flour was chosen:

> Why wheat? Because technologically, in Indonesia, the staple food that meets the criteria for fortification is wheat. Because rice is produced by millions of people. And again, we have been trying with rice, but it's very complicated. In Latin America, sugar. But for Indonesia, sugar is an unstable commodity. Production, price, imports. So we cannot work on that. Wheat is good because production is controlled, and now the consumption of wheat is going up, it reaches everybody—even the poor. So, if you make effective iron fortification [of wheat flour], this will reach the poor.[18]

Wheat flour also seemed like a good candidate because of the industry's close ties to the government. Since the Office of State Minister of Food Affairs was a spin-off of BULOG, the latter's close relationship with the milling industry was an asset. Indeed, a former staff member of the Office of State Minister of Food Affairs, who was identified as the prime mover for the project, had worked at BULOG before moving to the Office of State Minister of Food Affairs. He explained to me in December 2004 that wheat flour emerged as the best candidate partly because "Bogasari is an old friend of BULOG. Therefore, there were already long individual contacts so that we could work together."

In the meantime, nutrition researchers had gone ahead and conducted some research on wheat flour fortification. Scientists at the Center for Research and Development of Nutrition and Food (Puslitbang Gizi) conducted several studies (Komari and Hermana 1993; Muhilal, Murdiana, and Hermana 1997), and Bogasari did technical feasibility studies to confirm that there was no impact on taste, color, and cooking properties from fortification. By 1997, nutrition experts seemed to have solidly decided on wheat flour as a vehicle (Natakusuma 1998).

In June 1998, the government came close to finalizing the wheat flour fortification regulation. The Ministry of Health issued a decree on the mandatory wheat flour fortification program.[19] This regulation stipulated that all wheat flour be fortified with iron (60 ppm), zinc (30 ppm), thiamine (2.5 ppm), riboflavin (4 ppm), and folic acid (2 ppm), at a minimum. Deciding the details of fortification, nonetheless, took some more negotiation between researchers and the industry. When deciding on the iron level, Muhilal of the Center for Research and Development of Nutrition and Food initially argued that 50 percent of the Indonesian recommended daily allowance would be a good rule to adopt.

But industry rejected this as too much, and in response, Muhilal changed the recommendation to 25 percent of the RDA (Muhilal, Murdiana, and Hermana 1997), which would have been 60 ppm.[20] But the milling industry opposed this amount as well, and pushed for 50 ppm of iron. The experts compromised, and the final Indonesian National Standard became 50 ppm. Hence there was a slight change between the 1998 announcement and the 2001 SNI. The type of iron was another tricky question to be resolved. Researchers wanted to use ferrous sulfate and ferrous fumarate (Komari and Hermana 1993). However, the industry opposed this, citing the unattractive color of instant noodles if made with this fortified flour, based on an experiment conducted by Bogasari (Purnama 2002). Finally, the experts agreed to use another type of iron called elemental iron. The final SNI issued by the Ministry of Industry and Trade in 2001 was therefore a product of multiple compromises struck between experts and the industry.[21]

International organizations and bilateral donors were also keen on promoting fortification and eager to help the Indonesian program take off. Once wheat flour seemed to be the consensus, international organizations expedited the process. The major push for flour fortification came from USAID, UNICEF, CIDA, and ILSI (Maberly 2002). USAID donated through UNICEF the initial premix (iron, zinc, thiamine, riboflavin, folic acid) sufficient for 1 to 1.5 years in 1999.[22] This premix was distributed among Indonesia's mills according to their installed capacity (Purnama 2000). In 2001, CIDA provided a grant to UNICEF for assisting Indonesia's fortification project as well as 232,440 kilograms of premix, which was distributed to the four flour millers.

In sum, wheat flour fortification was the fruit of the existing technoscience network that had worked for some time to materialize mandatory fortification in Indonesia. The extension of this network was greatly facilitated by various factors. As we saw, the survey on anemia was cited by many informants as the critical event behind the eventual materialization of fortified wheat flour. International aspects also mattered significantly. The ascending profile of fortification in the global development community and the community's working relationships with Indonesian scientists and bureaucrats, along with their eagerness to showcase Indonesia as an exemplary case of mandatory fortification helped boost the morale of the network and expedite the process. And last, it was important that the new Office of State Minister of Food Affairs, with its useful connections to the milling industry, also sponsored the initiative. The network was able to frame fortification as falling between agriculture and health, an authentic "food issue," in a way that worked to get the sponsorship of the newly established ministry. The groundwork had already been laid so that the industry could take advantage of fortification. This time, they did not veto the plan and willingly followed the path charted by international and domestic experts.

Nutritionism and Its Blind Spots

If it was not solely due to the muscle of a powerful industry, why did wheat flour fortification happen despite its dubious cultural, economic, and health consequences? In search of an answer to this question, I have looked at two previous fortification attempts. Once it is situated in the longer history of fortification in Indonesia, it is clear that wheat flour fortification shares much with these two earlier fortification attempts, and what ties them together is nutritionism. In all three cases, nutritionism worked to narrow down the issue from malnutrition to a problem of "deficient" food and to the lack of particular nutrients. The equation of malnutrition with some form of dietary deficiency must not be naturalized, however. The etiology of vitamin A deficiency (in the case of MSG) and iron deficiency anemia (instant noodles and wheat flour) is quite complex, and there could be causes for these micronutrient deficiencies other than dietary deficiency, such as infection and other diseases. Nevertheless, there was a single-minded focus on food and its nutritional composition. It was deficient food that was blamed for micronutrient deficiencies. This ultimately justified the nutritional fixes whose sole mission was to add missing nutrients.

By reducing food to a collection of, or a vehicle for, measurable nutrients, nutritionism calculates food's worth in terms of the amount of nutrients it delivers. Such a microscopic view of food made it hard for involved experts to see cultural and social issues. For instance, one might have legitimately questioned the implications of making MSG and instant noodles into officially sanctioned healthy foods, since concerns had already been raised about cultural and health effects of these products. Similarly, once outside of the logic of nutritionism, one might question the use of crony capitalists under the Suharto regime as partners for public health policy. Yet nutritionism enabled experts to pretend that they were merely tackling technical problems rather than social problems. For them, politics and culture did not matter, because they were strictly dealing with nutrients. Furthermore, it is worth pointing out that this narrow attention to micronutrients sometimes did not even make medical sense. For instance, fortified instant noodles might be high in iron, but that does not reduce their sodium content. Yet as long as nutritionism singled out iron content as the signifier of food quality, fortified instant noodles could be legitimately promoted as healthy food.

Understanding the function of nutritionism helps to make sense of the apparent confidence that nutrition experts have in wheat flour as an ideal food policy tool. For virtually all of those I interviewed, it was not a case of business interests taking over a public health project. It was seen as a truly happy marriage of business and science without any guilt. The characteristics of the wheat flour industry—that it was monopolized, that it was making large profits, and that it had a close relationship with the regime—met with the approval of the experts, who saw this situation as providing ideal conditions for their pet project. In the nutrition experts' language, it meant that wheat flour satisfied the conditions of centralized production, ability to absorb additional cost, and easy enforcement of regulation. However, the flour milling industry might well be described alternatively as oligopolistic, protected, and politically well connected. Critical in constructing a deceptively simple and perhaps naive narrative was nutritionism, which connects wheat flour fortification with previous experiences with MSG and instant noodles.

We Know What You Need: Nutritionism and Science Governance

The story of fortification in Indonesia should serve as a critical reminder of the implications of nutritionism for science governance. One might assume that technical policymaking on nutrition and food is better left to experts and bureaucrats with technical expertise on the issue. However, among science and technology studies scholars and those involved in policymaking, there is a growing awareness that even technical policies and programs need to be founded on a democratic footing. Contrary to the classic portrayal of science rooted in positivism, science is never a neutral tool in policymaking. Science involves an exercise of judgment and is founded on implicit normative assumptions. The history of the discipline, conventions, and socialization also restrict science's frameworks and its approaches to any policy issue. Hence, increasing numbers of scholars have called for discarding the old expert-monopoly model of science in favor of the governance of science by citizens (Kleinman 2000).

A leading science and technology studies scholar, Sheila Jasanoff, has argued for science to strive for what she calls "technologies of humility" (2003). With this concept that emphasizes science's inherent political nature, she argues for more engagement between experts and citizens in which "citizens are encouraged to bring their knowledge and skills to bear on the resolution of common problems" (227). She points out that experts' humility in seeking voices from citizens is necessary, as science often fails to consider issues that fall outside the conventional framing of a particular discipline or that have to do with long-term consequences and differential exposure to risks and benefits of particular policy interventions or technologies.[23]

Nutritionism, however, exists in stark contrast with such humility about technoscience's role in food policy. Notice that citizen participation was almost entirely absent from the history of fortification looked at here. In all three cases of fortification attempts, there was very little direct or indirect input from regular Indonesians. There were several meetings that were dubbed "public" or "socialization," but in reality, most of them were attended only by bureaucrats and scientists. One might try to call the Widyakarya Nasional Pangan dan Gizi a place for public dialogue about malnutrition issues open to citizens, but it would be a stretch. Widyakarya was attended only by academics, policymakers, and some international invitees. We might imagine that there were "socialization meetings," which are regular fixtures in the Indonesian policymaking process and are supposed to increase public participation. Although I could not confirm the occurrence of such meetings before 1998, the ones I attended in 2004–5 only involved nutritionists, representatives of international organizations, and government officials. The KFI's opinion was taken as citizens' input, but as indicated earlier, all of its members are ex-government officials, nutrition researchers, or corporate representatives.

Instead of encouraging grassroots-based food policy dialogue, nutritionism effectively and significantly limits the stakeholders that are invited to take part in it. Key to this closure is the sense of certainty about the definitions of the problem and the needs that nutritionism offers. By reducing needs to nutritional components of food that are quantifiable and knowable by biochemical measurements, the statistics themselves become the authoritative account of "the food problem" and the needs of the people. It is not surprising, therefore, that all three cases of fortification in Indonesia crucially used survey data to make the case for fortification. Experts discussed fiercely the technical merits of survey data—whether it is representative of the population, what kind of measurements must be taken, which demographic groups need to be oversampled, and so on—but they took it for granted that they could rely on survey data to define the food problem and to understand the needs of people. The most valid and authoritative account of the needs of the people was, in their view, a survey with sufficient sample size, covering the whole nation, and conducted by the experts. It was for this reason that wheat flour fortification got a major boost when the national data on anemia became available. For experts, this data on the hemoglobin levels of pregnant women and children authoritatively fixed the shape of the problem and provided the indisputable evidence of the needs for fortification. Diagnosis and prescription were self-evident—hence, the people did not need to be heard.

Ironically, while nutritionism was instrumental in excluding regular citizens as relevant actors, it also worked to include the food industry as "experts" in food reform. By reducing food to a set of nutritional parameters, nutritionism refashions food into a mere carrier of macro- and micronutrients. This microscopic definition enables the business community to claim an expert status on food problems, particularly in the case of fortification, because the industry has expertise in the mechanics of adding nutrients at factories. Malnutrition became a manufacturing problem with scientists as experts on etiology and epidemiology and the industry as experts with practical know-how about fortification and consumers.

In a sense, nutritionism is highly productive. It is critical for fortification policymaking in its ability to weave both scientists and the industry into its fabric. Scientists could come up with the best fortification vehicle from their vantage point, but they still need to enroll business partners. Nutritionism laid the basis for the collaborative relationship between the scientific and industrial communities.

The stories of fortification in this chapter also show that nutritionism put industry and nutritional scientists on a par as experts. In all three fortification attempts, industry played a significant role in making or breaking the deal. MSG and instant noodles did not become official fortification policy due to industry opposition, whereas wheat flour did with industry's support. The equal expert status given to the industry was also seen in the negotiation over the level of iron fortification, in which the scientists' proposed iron standard for wheat flour was rejected by the industry and the industry recommendation became the final standard. The negotiation over the type of iron is another example. The final decision to use elemental iron, although opposed by scientists because it has much less bioavailability—about 50 percent that of ferrous sulfate, which was originally proposed, was made by the industry (Lynch 2005). This kind of back-and-forth between industry and scientists was considered necessary, not a distortion of science by economic interests.

The food crisis of 2007–8 is another reminder of the power given to private industry under nutritionism. The milling industry this time lobbied the government to lift the fortification requirement. This further illustrates that nutritionism might help to create a comfortable partnership with private corporations when they are willing, but it obfuscates the fact that corporations' loyalties lies not with the malnourished but ultimately with stockholders.

The case study of wheat flour also points to contradictions of neoliberalism. In examining neoliberal development policies of Asian developing countries, anthropologist Aihwa Ong observes that they "combine authoritarian and economic liberal features" and "are not neoliberal formations, but their insertion into the global economy has required selective adoption of neoliberal norms for managing populations in relation to corporate requirements" (2002, 236). The neoliberal orthodoxy articulates with locally specific conditions and historic contingencies, producing ambiguous results not easily captured by the framework of competitive markets and free trade. While fortification can be celebrated as a "public-private partnership," we should not conflate the global rise of neoliberal discourse and the pursuit of neoliberal strategies with what actually happens on the ground. Ironically, in the context of the late 1990s in Indonesia, fortification was used against trade liberalization and deregulation.

The paradox of nutritionism is that while it is instrumental in forging the critical alliance of powerful actors, corporate actors are included in the dialogue as experts although they are not accountable to the marginalized in society. It creates a comfortable space for the scientific and industry experts and authorizes them to define the problem and needs of the people without listening to them. In Indonesia the microscopic language of nutritionism shared by these experts naturalized the large role played by the corporate sector in fortification policies as well as the absence of discussion of issues such as the cultural inappropriateness of instant noodles and MSG fortification programs and the economic distributive effects of wheat flour fortification. Far from being "humble" in recognizing the need for citizen participation, nutritionism sets out to dictate the problem and needs of people, delegating this authority to limited experts.

6

SMART BABY FOOD: PARTICIPATING IN THE MARKET FROM THE CRADLE

It appears that it is practically impossible to supply enough iron from unfortified complementary foods to meet the iron requirements of infants....The situation appears to be similar for zinc at 6–8 months.

—World Health Organization, 1998

Only in the late 1990s did scientific evidence demonstrate that traditional homemade [foods], whatever the cost, could not meet infant and young children's micronutrient requirements, especially for iron and vitamin A. In order to meet infant and young children's requirements, they need fortified complementary foods that are only available commercially.

—Soekirman, 2005

Wandering through the maze of narrow streets that crisscross a Jakarta neighborhood, I finally reach Ibu Eti's place. Dilapidated and tilted, the shack looks like it is about to collapse. There are several plastic buckets outside and a man is doing laundry, squatting. I meet Ibu Eti and some of her five children. Ibu Eti's husband is the one who is doing laundry. He does it now because he lost his job. I ask what Ibu Eti does, and she hesitates a bit before saying that she begs on the street. When she became pregnant with the fifth child, she says, she gave up and started doing it because there was not enough money to get by. Like many women whom I have seen on the pedestrian overpass above Jakarta's chronically congested roads, she begs with the baby on her lap. Despite their poverty, she tries to use various commercial baby food products. She explains that although she cannot afford Dancow (formula by Nestlé), she has managed to buy Promina (weaning food by Indofood) to give to her kids and shows me the shiny package in the dark room where we are sitting.

Nutritionism brings a new visibility to women's and children's nutritional status. With the growing interests in hidden hunger, baby foods and foods for pregnant and lactating mothers have been subjected to increasing scrutiny as to micronutrient composition. Whether or not food for babies prepared by mothers

at home fulfills the micronutritional standards has become one of the important questions posed by policymakers and nutritional scientists. Governmental
and nonprofit organizations started fortified baby food programs, and people
like Ibu Eti have received, for instance, fortified cookies as part of an antihunger program. A growing number of nutritional studies suggest that the majority
of homemade baby foods are nutritionally "suboptimal," especially in terms of
micronutrients. From such a perspective, mothers like Ibu Eti, who buy fortified
baby food products, are "aware" mothers.[1] But how do we make sense of the
rise of such "smart" baby food for the poor and the official advocacy of them in
developing countries? In what ways do poor mothers respond to the scientization of baby food? What does it tell us about the politics of motherhood in the
contemporary global South?

Preparing food for one's family, particularly young children and babies, might
seem like a quintessentially private experience, a realm of love, care, and intimacy.
Yet the rise of smart baby food is the product of both scientific and nonscientific
diagnoses involving public health science, government nutritional policies, and
corporate estimates as well as women's own diagnosis of their needs and the
needs of their children. Not only mothers and babies but also those in science,
business, and the state have a stake in how babies are fed.

The story of smart baby food can be thought of as an aspect of the growing
control of women's intimate space by scientific expertise. The *scientization of
motherhood* refers to processes by which mothering practices have come to be
defined as scientific issues, resulting in a greater role for scientists and experts
(Chase and Rogers 2001).[2] Feminist historians have documented processes in
which problems related to mothering came to be seen as better addressed by
experts than mothers, requiring intervention through rationalized and science-
based regimens (Badinter 1981). What counts as good mothering is increasingly defined by medical and child-rearing experts. As a result, reproductive
issues such as contraception, pregnancy, and childbirth have become subject
to professional controls (Ehrenreich and English 1978: Margolis 1984), and
children are increasingly considered to be in need of expert instructions and
scientific products to be properly modern and civilized (Ladd-Taylor 1994).[3]
Analyses of contemporary motherhood discourses suggest too that the importance of turning to experts for advice continues in the contemporary ideologies
of motherhood that glorify "intensive mothering" (Hays 1996) and "sacrificial
motherhood" (O'Reilly 2004). Familiarity with scientific assessments of all sorts
of parenting practices still counts as an important requirement for being a good
mother today.

Because we are so accustomed to the image of developing countries as back-
ward and barely fulfilling basic needs, scientization might seem irrelevant in

developing countries, perhaps too fancy to be discussed in the context of the Third World. Yet mothers in developing countries are no less subject to pressure to achieve scientized motherhood. Interlinked fears of ignorance, poverty, underdevelopment, and negligence still make mothers suspect in the context of the contemporary developing world. Mirroring the colonial fear of "cultural contamination" (Stoler 1995, 72) from native mothers, Third World women's mothering practices create great anxiety in the minds of development experts and state bureaucrats. These mothers' cooking, feeding, and nursing constitute a high-stakes game for state and international development experts, and their dutiful compliance with expert instruction is important for development to take place.

The scientization of motherhood does not just devalue mothers' personal experiences and experiential knowledge vis-à-vis scientific and expert assessments and instructions. The privileging of scientific expert knowledge also accelerates the commodification of motherhood. As many observers of the scientization of motherhood have noticed, the corporate world has had a hand in molding the intimate space into not only a scientized but also a commodified space, offering various products and services to help translate expertise into consumption choices for mothers. Measured against scientific criteria, commercial products seem to fulfill the unmet needs of babies and children with tremendous accuracy and effectiveness. Helping to construct the superiority of the corporate offerings, the scientization of motherhood then subjects mothers to not only new languages of science but also new "choices" for consumption. Even the very poor are under such pressure. Ibu Eti, for instance, told me how much she wanted to buy brand-name formula. Agonizing over different products to pay for out of what she gets from begging on the street, and what to give up in buying them, she is in a situation that captures the striking contradiction of nutritionism in its capitalist incarnation and its impact on motherhood. In this chapter I situate the smart baby food phenomenon in Indonesia at the intersection of development discourse, transnational knowledge circulation, and global capitalism, and explore this complex interaction as it changes the meaning of motherhood in developing countries.

Ironically, what is sidelined in the world of scientized motherhood that is inhabited by scientists, policymakers, donors, and companies, is women like Ibu Eti who are actually trying to feed their children. Preparing food for one's family, particularly young children and babies, is an emotional and personal experience for many women, and the process of scientization has profoundly influenced the meaning of baby food for them. If science and market are increasingly singing the gospel of micronutrients, what do their target audiences have to say? What are the mothers' understandings of the food problem, and how do they relate to the scientized understanding of infant feeding?

After examining the expert discourses in the first half of the chapter, I move back to women's worlds by asking how *they* think about their baby food. I asked mothers in Jakarta's slum areas about their experiences of feeding their children. These interviews are important for my broader argument on nutritionism as well. So far, my narrative has focused on the discourses of experts and government bureaucrats, and the "beneficiaries" of their policies and programs have been on the periphery of the debates, often abstracted as numbers. This is, in part, by necessity. Two commodities that I examine in this book as representative of mandatory fortification (wheat flour) and biofortification (Golden Rice) have no direct connection to consumers. Wheat flour's fortification is perhaps rarely noticed by consumers, and Golden Rice was not yet marketed. By focusing on baby food, in this chapter I provide a rare glimpse into the consumer side of the story. The stories told by mothers further delineate the power and limits of nutritionism in defining the nature of the food problem in developing countries.

Making "Smart" Baby Food

Infants and children are considered a "vulnerable" population for micronutrient deficiencies, as for many other diseases. Global prevalence is estimated at 127 million preschool-aged children under five with vitamin A deficiency (West 2002), and 45.8 percent of children under five in Asia and 40.4 percent in Africa with iron deficiency anemia (UN ACC/SCN 1998). In addition to vitamin A, iron, and iodine, researchers are also discussing the possibility of widespread zinc deficiency (International Zinc Nutrition Consultative Group 2004).

There are various causes for babies' micronutrient deficiencies beyond simply "bad" food. Although micronutrient deficiencies might seem to be caused simply by foods deficient in micronutrients, their etiologies are actually far more complex. For instance, infection is an important cause of anemia, and hence hygiene improvement can be a policy prescription to prevent iron deficiency. However, as we have seen, it is food that is currently getting attention as the culprit in deficiency syndromes while other possibly relevant issues such as housing and water sanitation are sidelined.

Along with such focus on micronutrients in food, chemical analysis of baby food has become intense. In particular, what nutrition experts term "complementary food" (CF) and "supplementary food" (SF) for babies has become a target of micronutritional analysis. CF is defined as additional food provided to infants and young children (six to twenty-four months) to complement breast-feeding. SF denotes food provided to children or pregnant women in addition to their regular daily food.

The global expert recommendation regarding infant feeding is "exclusive breast-feeding" (feed only breast milk without any other liquid or food) for the first six months, as experts think breast milk fulfills all the nutrition requirements of the baby until that point. After six months, nutrition scientists recommend that babies get CF. Many parts of the world depend on homemade CF and SF, and scientists have begun to examine their micronutritional values. A number of studies have found that many of these are not up to standards in terms of micronutrients (Brown, Dewey, and Allen 1998). Therefore, commercial baby food has become an attractive option in the eyes of experts, as exemplified by these statements from WHO:

> It appears that it is practically impossible to supply enough iron from unfortified complementary foods to meet the iron requirements of infants....The situation appears to be similar for zinc at 6–8 months. (1998, 106-7)

> Rapid urbanization and changing social networks affect caregivers' ability to use freshly prepared home-grown foods. Centrally processed fortified foods, which can play an important role in ensuring adequate complementary diets, have been successfully promoted in various settings. Public-private partnerships can play an important role in making available nutritionally adequate low-cost processed foods. (2001b, 4)

Of course, there has been considerable tension among nutrition experts regarding recommending commercial fortified foods in the Third World. Expert recommendations often carefully include fresh fruits, meat, and vegetables as possibilities in addition to commercial fortified food. Yet the suboptimal micronutrient level of many traditional baby foods in developing countries has resulted in a growing emphasis on commercial alternatives.

The superiority of fortified baby food is also buttressed by constructing the inferiority of other nutritional interventions, such as nutrition education and supplement distribution. Fortification generated tremendous excitement among experts in part because they were frustrated with other micronutrient strategies, namely nutrition education and supplement distribution. Fortification seems better because it has less compliance problem. Due to this comparative advantage over other micronutrient strategies, fortified baby food has become a popular project for many international organizations and nongovernmental organizations. For instance, among the organizations with fortification missions discussed in chapter 3, many chose to fortify baby food. The International Life Sciences Institute started fortified CF as one of their focus projects in Southeast Asia. Global Alliance for Improved Nutrition (GAIN) started to provide fortified infant food in India and elsewhere.

Mirroring this international trend, Indonesia has similarly seen a fortification boom. Customarily, Indonesian mothers feed babies porridge made from rice and rice flour, bananas, papayas, beans, and vegetables (Komari 2000). Following research done elsewhere, Indonesian scientists started to examine such food in terms of its micronutrient makeup, and many concluded that these homemade foods for infants did not meet micronutrient requirements (Komari 2000; Ministry of Health 1999). Indonesian mothers' ability to prepare "good" baby food at home increasingly became suspect, and commercial baby food emerged as the ideal alternative.

If homemade food is the common icon of love and caring, new science paints a starkly different picture. Summarizing the newfound virtue of commercial baby food over homemade food, Indonesia's leading nutrition experts called for a "new paradigm of baby food" in a 2005 report:

> The new paradigm affects the common or existing concept of complementary food for infants and young children. The old paradigm stated that there was no difference between home-made complementary foods and commercial or factory-made complementary food. The new concept reveals the significant difference between the two complementary foods, especially in terms of micronutrient content and bioavailability. (Soekirman et al. 2005, 31)

Seen through the "new paradigm" of nutritional science, homemade baby food now stands as the icon of inferior quality and unenlightened feeding practices, an antiquated paradigm to be cast off for the celebration of the commercial alternatives.

Experts' shifting technical assessments led to material changes on the ground. Persuaded that smart baby food was the next big thing, the government and international organizations started fortification projects for babies and mothers in Indonesia. For instance, the World Food Programme started targeting mothers and infants in the distribution of fortified cookies and instant noodles. Other organizations, such as International Relief and Development, Land O' Lakes, and Helen Keller International started distributing fortified products to infants, mothers, and children. The Indonesian government also started a nutrition program called MP-ASI, the Indonesian abbreviation for "complementary food to mother's milk," using an instant baby porridge fortified with micronutrients. In 1999–2000, the government spent Rp 30.9 billion, about $3.4 million, for MP-ASI, and in 2003–4, Rp 120 billion, or about $13.3 million (Soekirman et al. 2005). This ongoing MP-ASI project has become a centerpiece of their nutrition program, constituting a significant bulk of the 2003-4 nutrition-related budget of the Indonesian government.

Who benefits from this "new paradigm"? What are the implications of these public endorsements of commercial baby food? While the government and non-profit institutions emphasize babies as the beneficiaries and the scientific virtues of these projects, the partner in this endeavor—the private sector—cannot be erased from the picture. The baby food market has moved beyond the Western capitalist states (Dunn 2004) and is growing at an impressive rate globally. Even in the developing world, the commercial baby food sector has a strong presence. The Indonesian baby food market epitomizes the spatial expansion of the baby food market and its increasingly powerful presence in the daily lives of families in developing countries. Indonesia's precise market size is unknown, but it was estimated in 2011 to be $136 million in annual sales (IRIN News 2011). Between 1997 and 2004, the market is said to have grown by approximately 90 percent (fig. 6.1; also see INSTATE Pty Ltd 2003; Indofood 2003). There are a variety of products in the Indonesian baby food market, but the market mainly consists of starter formula (0–6 months), follow-on formula (6–12 months), growing-up milk (1–10 years), milk for pregnant and lactating women, cereals/porridges, and special products for lactose low/free diets and for babies with low birth weight. The main products in the market in terms of volume are follow-on milk, growing-up milk, and cereals/porridges.

The majority of the companies in the business are foreign owned or multinationals, and major global corporations such as Nestlé and Nutricia compete in this market (fig. 6.2). For instance, the top player in the Indonesian baby food market is an Indonesian company called Sari Husada under Nutricia in the Netherlands (now a subsidiary of Danone). Essentially, two big players dominate the country's market—Sari Husada and Nestlé. Indonesia is an attractive market for these companies due to the large population of babies and mothers. As one of the businessmen whom I interviewed told me, it is a market of four million babies per year. Therefore, it is not surprising that Indonesia is one of the targeted growth markets (Madden 2003).

For these companies, the public and nonprofits' use of fortified baby food has been a financial boon. For instance, the government baby food program used baby food made by Gizindo-Kalbe Farma, which is a subsidiary of the food conglomerate, Indofood. The company was quite straightforward about the business benefits of the government food aid, as it reported to its investors in its annual report:

> Two years ago the timing of the resumption of aid-related contracts resulted in only 3 months of aid-related sales as the Government continued to support improved infant nutrition in Indonesia. In 2003, however, a full year of this business resulted in substantial growth in volumes and revenues. (Indofood 2003, 30)

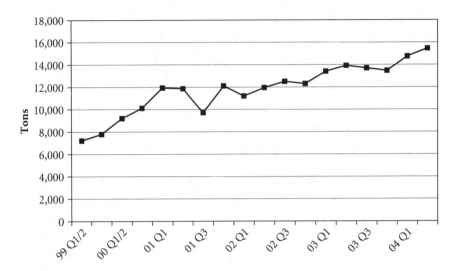

FIGURE 6.1. Indonesian baby food market, 1999–2004.
Source: APMB.

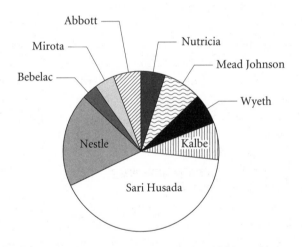

FIGURE 6.2. Indonesian baby food market share, 2003.

The public commitment to fortified baby food had benefits beyond such direct profits, however. Highlighting the longer-term investment value of public endorsement of commercial baby food, Indofood continued as follows:

> *One of our main challenges is the conversion of aid-related customers into the habit of buying our commercial brands as family incomes improve.* Aid agencies estimate a constant 30 million people below the poverty

line in Indonesia for the next 10 to 15 years and our capacity gives us
an advantage in securing more of this business in the future. (Indofood
2003, 30; my emphasis)

If nutritional scientists and bureaucrats thought that food aid would add micronutrients, the industry thought about adding future customers, as well as immediate "volumes and revenues." For the industry, the value of public endorsement
of fortification lies at multiple levels, first in the form of direct monetary gains
but also in the expected conversion from "food aid recipients" to "customers" via
the process of habituation of buying and eating commercial products.

The new paradigm of baby food has introduced mothers to not only the
new expert assessment of nutrition and food but also to a world in which the
private sector's calculation of growth and profitability figures prominently in
how they care for their children. On the one hand, smart baby food has been
touted by the national and global scientific network as a more enlightened
way for mothers to feed children. On the other, it has opened up the door of
financial opportunity for the global baby food industry, with its calculations of
profitability and business opportunities that are separate yet overlapping with
scientific judgments.

The Role of the Market

Ibu Lis is thirty-five years old and has three kids. She has an elementary school
education and lives with a husband who is a *becak* (cycle rickshaw) driver in
Jakarta. Like the house of Ibu Eti, her shack is dilapidated with no major fur-
niture inside the dark room except for a mattress. Although she does not have
what many might consider more basic necessities such as a decent bathroom
and kitchen, prominently sitting in the middle of the room are a TV and a DVD
player, which she got "on credit." When I ask her what she feeds her kids, she talks
about various commercial products such as Promina and Nestlé and says she
likes them because they are nutritious. I ask how she knows they are nutritious,
and she says it is written on the package and also points to the television and says
"advertisements."

Like Ibu Lis, other poor women whom I interviewed talked about the influence of advertisements on their feeding practices. If scientists and policymakers are emphasizing the value of micronutrients and commercial products as
superior alternatives to homemade baby food, how does it relate to the marketing
undertaken by the transnational baby food industry? What kind of story does it
tell consumers? How do these corporate actors talk, preach, and relate to mothers in developing countries? Here I look at the Indonesian baby food market and

how its marketing strategy has shifted. I sampled advertisements in a parenting magazine called *Ayahbunda*. It is a monthly parenting magazine written in Bahasa Indonesia. It has the longest history of the existing parenting magazines in the country, allowing for historical comparison of advertisements.[4] *Ayahbunda*'s readership is upper middle class, but from it one gets an idea of the advertisements by major baby food companies available in the Indonesian market, which are also seen in other media, such as TV, billboards, and in-store promotional materials that even the poor like Ibu Lis see everyday.

I found several salient themes in the 2005 issues of *Ayahbunda*. Most notable was their emphasis on nutrients and the benefits linked with a particular nutritional element. The most prevalent benefit featured both in text and images was the child's intellectual development. The main texts of ads emphasize intelligence as the biggest benefit of using the products. Images in the ads strongly emphasize the intellectual benefits by featuring children engaged in activities that seem to require brain power, such as children with objects associated with intelligence— complex toys (Dancow, 2005; Nutricia, 2005), computers (Procal Gold, 2005), and artifacts presumably made by the children themselves (Chil-kid, 2005; Enfagrow, 2005; Nutrilon, 2005).

In addition to the message of intellectual benefits associated with the use of the products, the 2005 ads tended to have a specific nutrient associated with each health benefit. Many established causality between a particular nutrient and a particular health benefit. For instance, the Milna biscuit ad claims DHA is "to help brain development," Prebio is "to increase body defense," and calcium is for "the development of strong bones and teeth." Gain Plus claims "three prime benefits" that include brain development from DHA, GLA, and taurine; bone and bone density from non-palm oil and calcium; and body resilience from sinbiotic. Triple Care similarly touts three benefits, including brain development from omega-3, -6, and -9; body resilience from beta-carotene, vitamins C, E, and B_6, and zinc; and improved digestion from fiber.

In order to identify the historical changes in the marketing strategies of baby food, I compared advertisements in 2005 with the oldest ones that I could find in *Ayahbunda*—from between 1979 and 1989 (summarized in table 6.1). Contrasted with the older ads, the emphasis on micronutrients and specific health benefits in later ads becomes very clear. Older advertisements were not silent on vitamins and minerals. Many of them mentioned vitamins and minerals in addition to proteins and calories. However, the major difference between the earlier ads and the 2005 ones is that while the overall message of the older ads focuses on children's growth and development, the more recent ones focus on nutritional components and their benefits, and on the complex engineering that is required to obtain optimal nutrition for babies.

In contrast to specific nutritional benefits in the more contemporary advertisements, the older ones emphasized children's growth as the central message. The image of a growing baby appears in various parts of the ads but figures most strongly in the main text: "healthy and happy growth" (Morinaga baby formula, 1979), "important for baby's healthy growth" (Vita rice flour, 1982), "nutrition for the growth of your children" (Proteina cookies, 1984), "healthy growth, growth to win" (Bendera formula, 1984), "Mil porridge for good growth" (Nestlé porridge, 1985), "Milna always cares about our babies' growth" (Milna porridge, 1986), "help baby's healthy and complete growth" (Farley porridge, 1986), "milk for our baby's growth and development" (Promil formula, 1986). The images in the ads reinforce this message. In addition to babies that are being fed, images of strong and active children are prevalent, including children holding giant stuffed animals (Nutrima, 1989), holding up training weights (Farley's, 1981), or being active in sports (Bendera, 1984; Sustagen, 1979). The older ads also emphasize additional benefits such as convenience, taste, and flavors.

This observation of the differences between earlier and contemporary ads is supported by a comparison of the same products from the same manufacturer over time. For instance, ads for Nestlé's Dancow milk in 1981 and 2005 reveal striking differences in their message over time. The ads in 1981 only talked about convenience and taste:

> Now Dancow Instant—delicious, from pure and fresh milk. Quickly— yes, in only 4 seconds Dancow Instant, rich in vitamins, can dissolve in cold water. Moreover, the taste is soft, white as snow. Soft and fresh— your children and family certainly will like it. No clumps. No waste. Dancow Instant—milk in 4 seconds.

In 2005, the advertisement for the same product emphasized the components and their specific health benefits:

> Dancow 1+, now more complete with DHA. Dancow 1+ is now not only giving protection with Prebio 1, which helps to protect digestive system, but is also complete with DHA, which is important for the brain. Dancow contains one of the highest amounts of DHA of all growth milk products.

The 2005 Dancow ad shows a picture of a girl whose brain is glowing as she plays with a complicated toy. The package now emphasizes nutrient components (DHA, Prebio 1, LA, ALA, and 26 vitamins and minerals), each with a specific composition. Each nutrient is checked with a √ symbol, inviting consumers to feel that all the necessary nutrients in the right amounts are in the product.

The text of the ad links a particular nutrient with a particular health benefit, such as Prebio 1 for digestion and DHA for brain development.

Thus nutritional makeup and its efficacy play an increasingly important role in more recent marketing. The 2005 ads portray the food for mothers' and children's needs as having a variety of nutrient components, each of which is specifically linked to a particular health benefit. The appeal to consumers is based on specific nutrient needs for specific health benefits, rather than the general "growth and health" appeal of the older advertisements. The important implication of this micronutrient emphasis is the necessity of expert intervention. The marketing messages construct the needs of babies as a complex amalgam of micronutrients, which then suggest the necessity of sophisticated engineering by experts. Therefore, they position professionals and experts, presumably at corporate laboratories, as superior providers of nutrition. Mothers are then framed as responsible for buying those products in order to be good mothers. For instance, an advertisement for EnfaMama highlights such mothers' responsibility by saying:

> For a new baby, I don't compromise. A healthy and smart baby is not just born. Mothers don't want to compromise during pregnancy and breast-feeding, because they play an important role. EnfaMama is complete with 65 mg of DHA and omega-6. DHA is clinically proven to help a fetus's brain development. Your sweetheart starts smart since birth, you don't want to compromise.

Peppered with numbers ("65 mg") and scientific names of nutrients ("DHA and omega-6"), this advertisement deploys the full force of scientific authority ("clinically proven") to imply that not buying the product is tantamount to an unacceptable "compromise" that results in suboptimal development of one's child.

Scholars have documented that the growth of nutritional science and the accumulation of findings in related disciplines have been accompanied by an influx of astute corporations that have translated science into purchasable goods and services (Apple 1987; 1996; Parkin 2006, chap. 6). Indeed, the history of baby food reflects the modernist march of capitalism to the drumbeat of nutritional science. The Indonesian case described here also attests to the continuous refinement of corporate marketing in using the latest scientific benefits to appeal to mothers and capitalizing on women's guilt and fear. The changing and growing demands of scientized feeding construct the industry as the best equipped to serve the needs of babies with their expertly configured, professionally produced, and scientifically endorsed "smart" products. It would be a grave "compromise," mothers are told, not to use these products.

TABLE 6.1   Comparison of baby food marketing strategies, 1979–2005

	1970S–1980S	2005
benefits of the product	helps growth (of body weight) and builds strong bodies	specific health benefits (brain growth, body defense against disease, digestion, etc.)
nutrient advertised	emphasis on protein, frequent mentioning of carbohydrate	DHA, vitamins, and other micro- and macronutrients
food purpose	energy for growth	delicately engineered to optimize bodily/mental functions
appeal to consumer	health	optimal health

Interviews with Women: "Needs" for Nutrition in Poverty

When the experts and corporations are sending messages that focus on the nutritional makeup of food and the functional advantages of nutrients delivered via commercial products, how do people interpret it? Merely describing advertisements is not sufficient because consumers are active participants in this commodified communication, and their agency has to be brought into the picture. In an effort to uncover the lost voice of women, albeit partially, I interviewed Indonesian women.[5]

I chose thirty-nine mothers in urban slum areas. I selected poor families for several reasons. First, poor women are the most frequent objects of interventions in the official and scientific discourses. It is poor women and their children that nutritional experts and policymakers consider as the primary beneficiaries of micronutrient policies. Second, at the same time, we could expect that poor people are the least likely users of the commercial products—particularly when the products are substitutes for their own breast milk, which is available for free. These are also the people who we tend to assume are the least likely to be exposed to nutritional science. Therefore, I wanted to see how these least likely consumers of nutrition messages were responding to the recommendations for commercial products. If they were buying baby food products despite their economic limitations, they must have had strong reasons. For the same reason, I focused on stay-at-home mothers rather than working mothers, since the latter had real reasons to use commercial products instead of breast-feeding. I also focused on the poor because of the obvious political and social implications of their dependence on industrial food.[6]

Wanting to see nutritionism from the perspective of women's lives and lived experiences, and being aware that a rigid questionnaire was likely to reflect my theoretical and cultural biases, I asked broad questions to elicit the women's own narratives. I typically started by talking about myself and said I wanted to hear about the interviewee's experience as an expert. I did not have kids then, so I would mention that I did not have children and thus I was there to learn from them. Then I asked about their families, food, cooking, and their feeding practices. I also asked their opinions about food aid, such as that from the WFP and the government.

When I asked them about the general food security situation, economic difficulty emerged as the biggest concern. Many women's husbands did not earn a stable income. Cash was always in short supply. Their living conditions were destitute. I went to many quarters that had just been under water from seasonal flooding that had left garbage and puddles of dirty water all over the place. When I asked them about the food situation, they tended to link it with this general condition of poverty.

> For food, we don't care about nutrition. What's important is there is protein, and kids become full. So the issue is different. For middle to upper class, they might think about nutrition and health. We don't think about health, nutrition, because nutrition is expensive. (Interviewee 146)

> For us, for people whose conditions are rather difficult, [we eat] whatever there is, whatever we can afford. Whatever is cheap. Mostly the problem is money. We don't think about nutrition. Whatever there is. Whatever is welcome. We don't think about nutrition. If we had more money, we would buy something nutritious. (Interviewee 168)

I was surprised, therefore, that even within this poverty, many women tried to buy commercial baby food products. Ibu Eti, who begs on streets, is one of them. The most popular products used were instant porridges for babies, such as Promina and Nestlé. Promina is a powdered rice flour produced by Indofood. Nestlé is a similar product produced by Nestlé. Some women did use formula milk and follow-up milk, or milk for pregnant mothers. Interviewees who did not use these products regularly had at least tried them. Those who did not use them regularly had had to stop because their babies did not seem to like the products or because the product was too expensive, but they emphasized that they had tried.

In parallel with the results from a variety of other surveys done in Indonesia, I found that most of the interviewees did not exclusively breast-feed for six months as recommended by the health authority. Most of them did breast-feed but started feeding non–breast milk at an earlier stage than recommended. Some of them fed their babies homemade porridge, bananas, or honey water. But many of them used commercial products.

When I asked why they bought commercial baby food products, interviewees provided complex reasons for their decisions, such as convenience, hygiene, and affordability. The most popular product type, the instant porridge (Nestlé, Promina, SGM) is considered convenient and easier to prepare than traditional porridge, which might take thirty minutes to prepare. The porridge is also affordable. A pack of 20 grams sells for Rp 1,000 and can be bought in the neighborhoods at small vendors without traveling far. You need only open a pack and stir in hot water to make it. Another reason given was hygiene. Food vendors sell porridge, but it is seen as potentially unhygienic. Some interviewees also said that health workers or midwives recommended the products or gave free samples, which could also be a motivation for continuing to use the commercial products.

Nutrition, however, was many interviewees' first reason for purchasing baby food products. As we have seen, many products feature nutrition benefits in their advertisements. The products that they used most—Promina and Nestlé—were similarly advertised with nutritional claims, such as that they contained "iron, iodine, high protein and EFA linoleat." Although many interviewees did not remember the exact nutrients that they were attracted to, they suggested that the nutritional value of these products held great importance for them.

> Products like that, Promina and Dancow, have vitamins. What's nutritious is Promina and Dancow. That milk is highly nutritious. The effect on that baby is good. That's Dancow. (Interviewee 144)

> [I buy them] for the child's brain. That is it. I want my kids to be smart. So I buy them although they are expensive. (Interviewee 145)

> I want the kids to be healthy. If you drink Dancow milk, [the child's] growth is fast. It's good. (Interviewee 139)

> They make kids big. And help nutrition. That has vitamins, DHA. (Interviewee 141)

> We want them because we see that the nutrition is better compared to regular food. (Interviewee 154)

> My husband thinks that our child is very small. So to help his appetite, we give him vitamins, [we] give this [Nestlé, Dancow]. He is so small, and to make him fat, to help his appetite, we try this milk. (Interviewee 164)

These women had gotten an impression of the nutritional superiority of these products mainly from TV advertisements, talking to other mothers, and seeing product packages. The role of television was particularly important, and most interviewees had some access to, if not ownership of, a television. Even in very impoverished households, like that of Ibu Lis, there was a TV. They also read the product packages, although not the details of the nutrition labels. They tended to point to catchy logos of "DHA," "iodine," and "Prebio" on the packages.

Does this mean that women are duped by the marketing strategies of corporations? Or does this mean that women's "needs" are met by the "smart" baby food? My interviews suggested a much more complex picture of women's experience of nutritionism. Three themes emerged from the interviews. These women were receiving contradictory nutrition information from advertisements and other kinds of knowledges (conflicting nutrition knowledges). They also believed that nutrition is expensive because of their awareness of baby food products, most of which are unaffordable to them (nutrition is expensive). Therefore, they want to at least try the products if they can afford it (try it if you can). I will discuss each theme in turn.

Conflicting Nutrition Knowledges

While all of the mothers and family members who I interviewed used baby food products, they did not forget to tell me that they were aware of noncommercial types of nutritional information. They emphasized that they knew advertisements were advertisements and that they took them with a grain of salt. Many of them also noted that they were aware of nutrition in natural, noncommercial food such as vegetables, fruits, and meat. For instance, when I asked them what they considered as nutritious food, many of them said vegetables, meat, eggs, and fruits. Several of them recited the old-time slogan of "four is healthy, five is perfect" (*empat sehat lima sempurna*) that recommends staple (carbohydrate), side dish (protein and fat), vegetables, fruit, and milk. This was the old government nutrition education slogan used for several decades. Breast milk had even more strong support from the mothers. All of them knew that breast milk is nutritious and best for infants babies, and no one denounced breast milk. Many of them had heard recommendations of breast milk from midwives and health volunteers.

On the other hand, they were exposed to many advertisements on television. Many said they received nutrition information from advertisements. One interviewee commented:

> I am a layperson. Do not know much. I get information from *posyandu* [community health posts] when weighing [babies]. And also on TV there are many advertisements. This product is good, that product is good. I don't know. From TV, that's it. (Interviewee 144)

As we saw, advertisements for baby food products often focused on fortified nutrients. Mothers therefore learned that omega-3 is good for the brain, beta-carotene is good for the body's resilience, or Prebio 1 is good for digestion primarily from TV ads. Suggesting the necessity to purchase products in order for people to get nutrients, advertisements shake mothers' confidence in regular food as a source of nutrition. Consequently, women are getting contradictory information and images of what nutritious food is.

Faced with contradictory information, many resolved the tension by arguing that baby food products are used only as a "side dish," an "addition," or a "variation" to "the normal food" or breast milk, which are recommended by official science. Many interviewees who used formula milk insisted that it was only to "add variation" or to "add to breast milk," rather than to replace breast milk. For instance, the mother who used infant formula and porridge from the time the baby was three months old said, "What's good is breast milk. That time [when babies are small], breast milk is better. It [formula milk] is only a side dish" (Interviewee 171).[7]

But, at the same time, some of the women had started to doubt the virtues of noncommercial alternatives in their particular conditions. For instance, interviewees knew about the medical recommendation for breast milk. Many of them repeated the phrase "breast milk is the best." However, some revealed deep concern for the quality of *their* breast milk. As many of them are poor, they believe that they don't eat enough nutritious food. They are told by health institutions to have confidence in breast milk as the perfect food for babies, yet they wonder whether that is only the case for the rich class of mothers.

> People in lower class actually…food is not nutritious. I eat little vegetables, little vitamins. Automatically, my breast milk for the baby is also not nutritious. So you need help by these [products]. (Interviewee 141)

> Sometimes, I am tired from work. Sometimes I don't eat too much, because I am busy. Then breast milk automatically decreases. If I am tired, breast milk is not enough. So I add SGM. (Interviewee 174)

Furthermore, health institutions can give contradictory messages about where people can get nutrition. Although many have heard that natural food can give nutrition and vitamins, the understanding of the term, "vitamin" is rather medicalized, strongly associated with supplements. Vitamin A distribution for children under five was started in the 1970s by the Indonesian government. Given the long history of the program, it might not be surprising that interviewees tended to equate the word "vitamin" with vitamin A capsules that they received twice a year for the children. When I asked "from what do you get vitamins?" a frequent response was "from *posyandu* and *puskesmas*."[8] Another frequent answer to the above question on the source of vitamins was commercial vitamin syrups such as Biolysine and Sakatonik. The women explained that they were advised to buy these vitamin syrups when their children were sick or underweight to help increase appetite.

On the one hand, women knew that breast milk was good and vegetables and other regular foods were nutritious. On the other hand, they had a medicalized image of vitamins and thought micronutrients had to be supplemented as specialized products. Women's understanding of nutrition is hence quite conflicted: while the former knowledge might affirm that commercial food is unnecessary, the latter undermines such understanding.

Nutrition Is Expensive

When asked about their general food security situation, the interviewees said that economic difficulty was the biggest obstacle in obtaining nutritious food for their families. They believed that they could not afford nutritious food in general and expressed feelings of deprivation. Although there are affordable foods that are nutritious, such as vegetables and tempeh, and they are accessible and sold in their neighborhoods, many women tended to link "nutrition" with something they could not afford.

> [The problem of getting nutritious food is the] economic difficulty. We want to buy nutritious food, but nutritious food is usually supermarket price. Sometimes we go to supermarket, but prices are high. (Interviewee 147)

> If I had more money, I would give my children food with vitamins. The nutritious ones. I mean, we don't have enough of it. We get milk only sometimes. I don't have much money. I buy milk, and that's it. Very limited. I mean, not much money.

> If you don't have money, you cannot buy anything. If I had money, I would buy food with vitamins, nutritious food. (Interviewee 144)

This theme, that nutrition is expensive, emerged perhaps because there was an increasingly strong sense that in order to "get" nutrition, you had to buy commercial products. Exposed to many advertisements, the women shared the sense that nutrition and vitamins are something rarely affordable to the poor. In the words of one woman, "The obstacle is money. If there is money, buy [nutritious food]. If there is no money, don't buy [it]" (Interviewee 169).

Nutritionism seems to accelerate their frustration by implicitly sending the message that in order to feed kids well, they have to purchase products. One needs a lot of cash to get nutrition because it is equated with buying nutrients. Echoing the message in advertisements, the ability to be a good mother is dictated by one's cash flow.

Try It If You Can

Women told me that they knew that regular foods like vegetables and fruits were nutritious. However, commercial nutrition information made them think it was necessary to buy products. Yet many products are beyond their purchasing power. This creates a sense of deprivation and paradoxically prompts a particularly strong desire to at least try products when their price seems more affordable. It was surprising to learn that many of these women knew a variety of products on the market, although they themselves could not afford them. Many of them could recite many product names in each category. For instance, if I asked them about cookies, they could talk about Milna and Sun. They knew Laktamil was for pregnant women. They knew many brand names of formula for babies as well. Furthermore, many of them knew the relative prices of products, telling me that product A is more expensive than product B, and that B is about the same as C. It was obvious that they knew of the existence of many product choices out there and carefully compared prices and affordability, eyeing the possible purchase.

Since many of the products are prohibitively expensive for the poor, the women's eagerness to participate in the market was strong when there were affordable products. A good example was the prevalent use of instant porridges like Promina and Nestlé. Many interviewees who used these instant porridges said that what they actually wanted to try was formula milk. However, formula milk tended to be more expensive than porridge. Porridges were sold at Rp 500–1,000 per sachet, while formula milk could be more than Rp 20,000 for a box, and it was not sold in a smaller quantity. For instance, one interviewee only used Nestlé and Promina porridge and Sun cookies. She had tried SGM and Sustagen formula milk but said she stopped because they were too expensive. Noting that these products "make children smart and are good for brains," she said those products "were good" but "the price was also expensive." Another interviewee, who had a two-year-old child, wanted to give the child formula milk but thought that the price was too high. She explained, "When fed with these products, children get fat fast. And growth is fast. Intelligence is good. If you drink this milk, you become smart. But we cannot afford it" (Interviewee 139). So she decided to at least buy the more affordable instant porridges.

Another interviewee put it this way:

> First is from curiosity. First was curiosity. Second, milk adds nutrition for children in addition to vegetables. I want to try it. Also at school they say that milk is necessary to make kids' IQ high. There is an IQ test, and we buy milk to increase the score. Just want to try. (Interviewee173)

Explaining that she just wanted to try because of benefits they saw in ads, the interviewee epitomizes many mothers' desire to buy if possible.

In summary, the interview and advertisement analyses provided a window into a complex world of feeding that mothers are faced with. The food industry markets their products as highly nutritious and functionally beneficial to children, often implying a responsibility and obligation to buy them to be a good mother. Interviews with the mothers also showed that advertisements created a deep ambivalence about breast milk and the allure of commercial products. Emphasizing the necessity for micronutrients in expertly manufactured products, ads seem to have created confusing terrain as to mothers' understanding of how and from where to get "good" food. Experiences with medical institutions and government programs reinforced the association of nutrition with commercial products. Women tended to lose their confidence in their breast milk, worrying about poverty's impact on it, which increased their desire to buy these commercial products. Catchy ads on nutrients and their benefits have thus created the strong yearning to buy commercial fortified products. Despite their poverty, many mothers at least try to feed these products when the situation allows.

Experts are increasingly worried about the micronutrient status of children and have come up with specific instructions for mothers to fulfill the micronutrient standards. A woman is told to breast-feed and not to add anything until her child is six months old. Yet after six months, she needs to buy properly fortified products because she probably cannot cook micronutrient-rich food for her child. Although this instruction might make perfect sense from scientists' point of view, it would be easy for anyone, let alone poor mothers, to fail to follow it. As any mother who has nursed her baby knows, breast-feeding is difficult to manage and physically taxing. In addition, poor women have reasons to think that their breast milk might not be enough for the baby. At the same time, commercial products tout great nutritional benefits that a woman might be tempted to try if she has an opportunity. In addition, nutritional claims are abundant in processed food products, making it difficult to distinguish what is properly fortified and what isn't. Furthermore, while women rarely encounter health workers who reinforce correct instructions, they are bombarded with confusing advertisements.

The expert discourse tends to hold mothers accountable for failing to follow scientific guidance. Experts lament that mothers do not breast-feed exclusively, introduce non–breast milk too early, buy wrong products, or cook food that is insufficient in micronutrients. But as the interviews showed, women are desperate to feed their children well. It is just difficult to do it for mothers when they are so impoverished. Many mothers pointed out poverty as the main concern even when I was asking about food and nutrition. Ibu Eti summarized the food problem: "For me, the problem is economy." Yet in the nutritional discourse of experts, it is rarely poverty that is the focus. Rather, it is nutrients and mothers who fail to deliver properly to their children. Shifting the focus from poverty and marginality to the matter of micronutrients, they effectively (albeit subtly) frame mothers as the source of the problem. Yet ironically, while the food industry floods women with advertisements that effectively build up the allure of their products by working on mothers' insecurity and desire, they are not problematized by the scientific experts who are primarily concerned with micronutrient, not a broader cultural and social environment in which women make their feeding decisions. Instead, the industry emerges as the significant partner in combating hunger and malnutrition because of its expertise in adding nutrients during processing.

From Villains to Saviors: Changing Politics of Baby Food

Several decades ago there was wide circulation of news stories about formula milk causing declining breast-feeding in developing countries. In the 1960s, the distribution and marketing of baby food and infant formula in developing countries by global food manufacturers invited much criticism from nongovernmental organizations, inspiring a global social mobilization against them. Corporations like Nestlé were called "baby killer," and they became the targets of transnational protest. Social movements pointed out that these multinational corporations aggressively marketed their products to developing countries with ethically questionable methods (Baumslag and Michels 1995, 154). They marketed formula where the necessary infrastructure to use them safely was lacking. Many mothers had to rely on dirty water to dissolve infant formula, making babies sick. The marketers of these products were often dressed like medical professionals, emphasizing the appeal of modern science to consumers. In some cases, medical professionals were paid to sell these products to their clients at hospitals. Corporate advertisements often compared breast milk and formula milk, falsely claiming the superiority of the latter over the former (Baumslag and Michels 1995; also see IBFAN website at http://www.ibfan.org/fact-nestle.html). By the early 1980s, there emerged a strong international movement against formula and baby food manufacturers, with the founding of organizations such as the International Baby Food Action Network (IBFAN) and the International Nestlé Boycott Committee. In response, WHO and UNICEF sponsored a codification, the International Code of Marketing of Breast-Milk Substitutes, which was approved by the World Health Assembly in 1981 (Sikkink 1986). The code attempted to regulate the marketing of formula and to promote breast-feeding by prohibiting, for instance, the use of formula company employees in prenatal education or the use of incentives to sell formula products.

Set in this historical perspective, what is impressive about the current "smart" baby foods is not so much the different nutrients added to them but the different cultural imagery. Unlike in the earlier period, when public health advocates were at war with global capital, smart baby food now enjoys the former's endorsement. This is a critical shift in the contentious politics of global food. The multinational corporations are now expected to contribute to, rather than undermine, the public health objective of infants' and children's health. As a publicly sanctioned hidden hunger strategy, smart baby food has achieved a subtle but profound change in the image of baby food manufacturers.[9]

Yet the ethics of baby food and formula marketing in developing countries is not only a concern of the past. Despite the establishment of the International Code of Marketing, for many breast-feeding advocates, the battle against baby food manufacturers in many developing countries is not over. Breast-feeding is far from optimally practiced, and companies have found many loopholes in regulations (IBFAN 2004). Even today, one of the most active campaigners, IBFAN, issues papers full of reports of violations of the code by corporations in developing countries.

Indonesia's politics of baby food resembles trends elsewhere. When the "baby killer" scandal broke out in the 1970s, the Indonesian government responded to the global concern and adopted a series of regulations to curtail infant formula marketing. In 1975, Indonesia's Ministry of Health issued regulations prohibiting advertisements for infant formula in maternity centers, hospitals, and other health service outlets. After the International Code of Marketing was adopted, the government issued a regulation in 1981 that prohibited television advertisements for infant formula (Office of the Minister of State for the Role of Women 1990) as well as several decrees to regulate labeling and promotion of breast-milk substitutes.[10] A regulation on advertisement and labeling was reissued in 1999,[11] which banned the promotion of baby food in mass media.

Despite the existence of these regulations, their actual enforcement has been quite limited (Utomo 2000). Marketing tactics that are banned by the code and other domestic laws are rampant in practice. Many observers have attributed the growth in sales of baby food in the past several years in Indonesia to aggressive marketing tactics (see, e.g., reports in Jakarta Post 2001a, 2001b, 2001c).[12] My interviews with experts also confirmed that the violations are widespread, with the government being unable to crack down on them due to lack of resources (see also BKPP-ASI and YASIA 2003; Utomo 2000). Some even said nostalgically that the situation was better during the New Order when the government was stronger. It is a sad irony in food politics that an authoritarian regime might be better able to enforce compliance with food regulations.

Moreover, various studies in Indonesia have indicated that the rate of breast-feeding is less than ideal. Many Indonesian mothers are not following the WHO/UNICEF guidelines of six months of exclusive breast-feeding and are introducing solid food too early or not breast-feeding at all. For instance, the National Household Survey in 2001 (Government of Indonesia 2001) shows that only 47.5 percent of newborn to three-month-old babies and 14.2 percent of four- to five-month-old babies were exclusively breast-fed. Another government source, the Demographic and Health Survey, showed similar numbers. A report by a NGO has found that only 37–41 percent of children less than two months old were exclusively breast-fed (de Pee et al. 2002). A smaller interview survey found that only 25 percent of mothers did exclusive breast-feeding for the first six months (BKPP-ASI and YASIA 2003). I do not intend to assess which of these data are more accurate. What is important here is that these statistics indicate that the reality is far from what is considered ideal. A number of health and breast-feeding experts in Indonesia whom I interviewed thought that even these surveys are overestimating the amount of breast-feeding.

If mothers are not following the six month exclusive breast-feeding rule, what are they doing? The results from the surveys indicate that many babies are fed with food and liquids too early. Many mothers introduced complementary foods before the recommended age of six months. To be sure, the introduction of food and liquid at an early stage of life can be seen as a long-standing custom (Nain and Maspaitella 1973; Hull 1979; Arnelia and Muljati 1993). Deciding that commercial products are responsible for the low rate of breast-feeding might seem premature. The growing trend, however, is that many mothers are using infant formula and commercial baby food. Helen Keller International reported in 2002 that many babies were fed with commercial instant baby food prematurely. Studies in other countries that analyze the reasons for declining breast-feeding practices also point to the influence of advertisements for baby food (see, e.g., Igun 1982).

Set against the continuing struggle over breast-feeding, the smart baby food endorsement from experts seems puzzling and profoundly contradictory. On the one hand, the experts have been concerned with declining breast-feeding and have tried to institutionalize various measures to convince women that commercial products are no better than breast milk. On the other hand, a newer discourse emphasizes the superiority of commercial products in terms of nutritional contents. In this confusing terrain of scientized motherhood in developing countries, it is the corporations that surely stand to win. Not only do corporations skim profits from public procurements, they can also proclaim their positive role in the health of babies in developing countries, refashioning themselves as the saviors, rather than killers, of babies in the Third World.

The Nutritionalized Self

Anthropologists have found that food often produces anxiety. Violating the cultural boundaries between pure and impure, nature and culture, human and inhuman, and life and death, food is full of ambivalence and contradictions (Levi-Strauss 1983; Douglas 1966). While one might expect that the "rational" perspective of nutritional science might reduce such anxiety, historians of food have found that this is not the case. As Harvey Levenstein (2003, 256) notes, "generalized anxiety" about food and nutrition is pervasive, and often aggravated by fluctuating and contradictory nutritional messages that frame food as a health risk.[13]

Gyorgy Scrinis's (2008) concept of the "nutritionalized self" echoes such a paradoxical increase in concern about food—even with or rather because of nutritional science. With this concept, he discusses the thrust for self-regulation and monitoring imposed by nutritional science. The concept describes eloquently how people have changed their relationship with food. We are perpetually worrying about the adequacy of our vitamin intake, taking multivitamins, jumping into new superfoods, and following news reports about scientific findings on micronutrients. The rise of nutritionism is accompanied by consumers who scan food from a nutritional perspective, avidly consume nutritional information, and modify their food choices according to the nutritional characteristics of products. It is not only the government, scientists, and corporations that have come to embrace nutrition as the primary parameter for food, eating, and feeding. Nutritionism has taken hold of the popular imagination, and the pervasiveness of the language of nutritionism in every corner of society, and in one's sense of self, is quite remarkable.

The concept of the nutritionalized self is useful in understanding the growing power of a nutritional perspective in shaping people's subjectivity in relation to food, but it would be misleading if it implies that all individuals are held responsible for nutritional well-being in the same way. This is not the case. Women—particularly as present and future mothers—are under more stringent scrutiny. Their cultural standing as mothers makes it ever more difficult for women to escape scientific scrutiny. In fact, nutritionism intersects with the modern ideology of motherhood, which prescribes that a mother must follow expert advice in order to count as a good mother. The theorizations of both Andrea O'Reilly (2004), on the ideology of "sacrificial motherhood," and of Sharon Hays (1996), on the ideology of "intensive mothering," acknowledge the increasing power of scientific knowledge and expertise in defining a culturally acceptable model of motherhood. For instance, O'Reilly notes that under the ideology of sacrificial motherhood, mothers need to be guided by expert instruction and scientific knowledge.[14] Good mothers are supposed to seek scientific wisdom on the psychological, physical, and cognitive development of their children. Nutritionism further accelerates such demands on women if they are to count as "good mothers."[15] Indonesian mothers whom I interviewed showed a similar pervasive censoring of self through the nutritional perspective. Indeed, what struck me during the interviews was that the mothers felt compelled to emphasize their knowledge of nutritional science. Mothers repeatedly deployed words like "vitamin," "protein," and "DHA" and recited the official nutritional slogans. Perhaps they did not understand scientific definitions, but they understood well the social and cultural value of nutritional jargon as indicating enlightened motherhood. In showing their familiarity with nutritional terms, they tried to impress on me that they met the requirements of good mothers.

It is precisely because of this gendered requirement that many corporations could use nutritional claims implicitly and explicitly targeted at mothers, capitalizing on their feelings of guilt, insecurity, and anxiety. As historian Katherine Parkin (2006) has shown, the vast majority of the food industry's advertisements have targeted women, particularly mothers, as the audience. Ads construct mothers as being in charge of nurturing and feeding children and family. It is motherhood, rather than fatherhood, that has been required to be scientized (Apple 2006) and commodified (Rothman 1989; Taylor, Layne, and Wozniak 2004; Paxson 2004). In the capitalist food system, the family nutrition is a mother's task that she should properly fulfill via the market.

The food industry is not the only actor that subjects mothers to strong pressure and tries to hold them accountable for the right purchasing decisions. Mothers' nutritional competence is also a concern for the nation-state. The nutritionalized self has a crucial link to what Nikolas Rose (2007, 24) calls "biological citizenship," in which "individuals themselves must exercise biological prudence, for their own sake, that of their families, that of their own lineage, and that of their nation as a whole". In being critically linked with neoliberalization, citizens' health and well-being have come to be matters of private accountability, with citizens refashioned as "responsible consumers" who must self-monitor their own bodily conduct. The concept of biological citizenship further mandates that citizens aspire to biological self-improvement in order to enhance the vitality of the nation for the purpose of "development." This requires women to be particularly prudent, because they are accountable not only for their own biological performance but also that of the future citizens. As Paxson (2004, 211) notes in the case of modern Greece, the government urges women to be proper "maternal citizens," to "fully achieve 'their biological mission' as women and to reproduce for the nation."

Indeed, in the everyday lives of the Indonesia women I interviewed, state policing regarding child nutrition seemed everywhere. For instance, children's growth is monitored at monthly weighing session at *posyandu*, and if the children are underweight, local health workers mark the mother as the "mother of a malnourished child." Such women might receive several visits by senior local women who act as local health workers and who will lecture them about nutrition and proper feeding. Local neighborhood festivals often have a little weighing table for babies, and officials nod approvingly if a baby is ahead of the official growth curve. The nutritional life of mothers has both economic and political values, and they are closely monitored and regulated.[16] Mothers are therefore constantly pressured to be nutritionally proper by both the state and the food industry.

Exposure to and familiarity with nutritional knowledge does not have to be negative; it can be empowering and productive. New knowledge about nutrition does excite people, and the "nutritionalized self" could be a site of desire and satisfaction. I have found myself taking a less-guilty pleasure in eating chocolate once I learned it is now considered good for you (antioxidants). Nor could I resist the temptation of buying a bottle of Vitamin Water. Several times I tried to feed my children DHA-added formula milk. I have found that navigating the market aisle that has smart foods with the anticipation of improving myself or my family could be fun and gratifying. In her discussion of scientific motherhood in the United States, Rima Apple (2006) similarly documents that many contemporary American women enthusiastically seek expert information to complement the decreasing amount of information obtained from family, sometimes forming a kind of "partnership" with experts. Scientific motherhood could also be strategically deployed by mothers to enhance social status and achieve upward social mobility, especially in minority communities (Litt 2000).

The ability to comply with scientific and expert advice, particularly when it is tightly linked with commercial opportunism, is highly class stratified.[17] The nutritionalized self is empowering and exciting only if one can afford promoted products, can seek more information if necessary, and can choose to withdraw from the nutritional mode. But my interviews with Indonesian women show a very different picture. Painful, is the word that I felt during my interviews with mothers. These women were fascinated by baby food products and the magic of vitamins. But as I was sitting in a dark, small room in a dilapidated shack or in a dark corridor running through a slum and smelling the dirty river or sludge next door, I could not ignore the profound gap between the mothers in the ads and the women I was talking with. The reality in which they found themselves could not provide them with the pleasures of nutritionism, and they knew it too well. Furthermore, imposed as nutritionism is by powerful actors including experts, the state, and the industry, there is no voluntary exit from it for these women.

The nutritionalized self opens up a difficult moral terrain for many mothers. Like Rayna Rapp's (1999) "moral pioneer," who confronts the new challenges of reproductive science and prenatal technologies, mothers have to navigate complex fields of new lines of products, confusing expert advice, personal and intimate assessment of their and their family's needs, and pressure to be good "maternal citizens." The nutritionalized self is a disposition riddled with contradictions and ambivalence. The obligation of women to take care of their bodies in relation to nutritional science, citizenship obligations, and consumer choices puts them to the task of learning new languages and calculations. At the same time, they often find it impossible to follow the dicta of nutritionalized motherhood due to their poverty and social marginality. Nutritionism, for these impoverished women on the periphery of the global economy, represents the extension of capitalist, state, and scientific control, not the extension of autonomy.

By discussing the creation of the nutritional self and framing nutritionism as a part of biopower, I am stepping into a fiercely contested theoretical arena. Feminists have long criticized Foucault for projecting a total hegemony of biopower and not recognizing the space for agency, resistance, and social change.[18] The projection of a totalizing power over subjectivity is problematic for feminists, who are committed to the possibility of empowerment and social justice. Nonetheless, it is important to realize that Foucault himself recognized that biopower is a project, often incomplete, contested, and always in the making. Foucault argued that subjects of biopower are not completely passive and that biopower does not deprive them of the capacity to reflect on their situations and act differently (Sawicki 1991). Empirical examinations of biopower also demonstrate the complex picture of subjectivity of subjects under biopower. In her analysis of American motherhood, Apple (2006) found that mothers in the 1990s started to criticize the expert-driven child-rearing dogma, although they were simultaneously profoundly influenced by it. Feminist theories on medicalization similarly have provided examples that challenged the image of women as passive victims of hegemonic medical discourses, while still noting the tremendous influence of medical discourses (Grant 1998; Lock and Kaufert 1998). Biopower may be "experienced as enabling, or as providing a resource which can be used as a defense against other forms of power" (Lock and Kaufert 1998, 7). In her analysis of scientific motherhood, Jacquelyn Litt (2000) argues that we need to understand mothers not as passive victims but as "agents that encounter and give meaning" (62) to scientized discourses. We have also seen subversive readings of modern nutritional claims by Indonesian women. To portray the disciplining power of nutritional science as omnipotent is to betray the awareness of the women I interviewed about their healthy suspicions about corporate marketing tactics and the incompleteness of the nutritional terminology. The women's narratives show that they were not simply brainwashed but rather that they have pragmatically embraced, resisted, and modified the dominant nutritional discourse.[19] Like some other feminist scholars who have theorized biopower, my contention is not to describe a complete takeover by biopower but to recognize its specific function. I do so with the belief that it then can be the baseline for emancipatory action. The project of making a docile subject is never a finished one. Science- and medicine-based discourses come with possibilities of both consent and dissent.

7

CREATING NEEDS FOR GOLDEN RICE

This rice could save a million kids a year.

—On the cover of *Time*, July 31, 2000

"World Top Economists Say Biofortification One of Top Five Solutions to Global Challenges" was the triumphant headline of a 2008 press release by HarvestPlus, an organization promoting biofortification. At what was called the Copenhagen Consensus Conference, distinguished economists were invited to think about major challenges facing the world. Asked to prioritize global issues, they pinpointed HIV/AIDS, global warming, and terrorism and others along with malnutrition, which they indicated had one of the "most effective solutions" in biofortification.[1] Nonexistent fifteen years earlier, biofortification had come to enjoy widespread recognition among development practitioners by 2008.

Biofortification uses plant-breeding technologies to develop food crops that are rich in micronutrients. Conventional fortification adds minerals and vitamins to a particular food vehicle during processing, before purchase and consumption. In contrast, biofortification reconfigures the plant itself through biotechnology or conventional breeding so that it is more nutritious. An often-cited example of a biofortified crop is Golden Rice—rice with daffodil genes inserted into it so that its endosperm contains beta-carotene.[2]

Until the late twentieth century, biofortification was not among the common policy tools for combating micronutrient deficiencies. The conventional package of "micronutrient strategies" in many nutritional science textbooks and policy documents usually included food fortification, nutrition education, and vitamin supplement dissemination. Indeed, the birth of the concept of biofortification can be traced to the mid-1990s, and the term never appeared in earlier nutritional science or crop-breeding literature.[3] Within less than a decade, the concept of biofortification became well known among international organizations, food policy experts, and nutritionists, and it attracted significant support. How did biofortification achieve such prominence on the radar screen of mainstream development practitioners? What accounts for its success in building a technoscientific network, carving out a niche among the weapons in the global fight against hidden hunger?

To answer these questions, we need to situate the story in a broader global politics of development, technology, and profit making than in a purely technical assessment. Although often celebrated as a major scientific breakthrough, biofortification is impressive more because of its expansive network of supporters than because of its actual nutritional efficacy. Biofortification's support network includes international agricultural researchers, bilateral/multilateral donors, philanthropic organizations, as well as biotechnology researchers and the life sciences industry. Its proponents have also effectively mobilized a moralized discourse to make it a part of the much-awaited "gene revolution" that will bring biotechnology to developing countries. In this chapter I analyze the social power of this particular "solution" to global malnutrition as a performance on a historically situated stage that elegantly weaves diverse interests, deploys emotionally powerful imageries, and claims moral superiority.

Despite the discursive success of biofortification, and in particular, of Golden Rice, such products have not actually been used by the poor in the developing countries. Here I analyze the case of Golden Rice in Indonesia. Interrupting the unquestioned celebration of Golden Rice, I consider the limited viewpoint of nutritionism, which has rendered many social and cultural issues invisible. While Golden Rice proponents use the poor in the global South as their moral foundation, the poor and the malnourished have not had a say in defining the needs to be addressed through Golden Rice research other than their representation in nutritional status studies. Like the hypervisibility coupled with absence of women, the "poor" has been constructed as a generic category for a target population. The Indonesian case study points to the discursive power of Golden Rice and its limitations.

Breeding for Nutrition

The promoters of biofortification tend to present it as a novel undertaking that combines agriculture and nutrition. It is true that, generally, agricultural research and development after World War II emphasized crop yields and income increase rather than nutrition as research objectives. The Green Revolution was resolutely grounded in this way of thinking, so that productivity became the most valued standard for measuring the success of agricultural research (Mebrahtu, Pelletier, and Pinstrup-Andersen 1995). However, in the 1970s, agricultural researchers briefly developed nutritional objectives and sought to breed for better nutrition for the Third World poor. This early movement by scientists to raise the issue of nutritional quality had been triggered by a realization that the Green Revolution might have increased agricultural yields but, ironically, had not necessarily improved the nutritional situation (Dewey 1979; Fleuret and Fleuret 1980). One of the criticisms of the Green Revolution was that the quality of diet for the poor did not improve, or that it even actually deteriorated, due to the new agricultural practices (Miller 1977; Manning 1988).[4] For instance, critics contended that pesticide runoff killed fish, destroying a precious protein source for the poor (Cleaver 1972). The iron content of the Asian diet was reported to have decreased due to the Green Revolution, as it replaced many of the foods based on beans and peas with cereals (Haddad 1999).

In response, the Consultative Group on International Agricultural Research programs started to examine nutrition as a potential research objective. There was, according to one observer, a "significant evolution in the perception of the ways in which agriculture, and therefore agricultural research, could influence nutrition" (Omawale 1984, 273), and affiliated centers started nutritional programs.[5] A nutrition focus in this period, however, did not mean a focus on micronutrients, but more usually a focus on protein (Bressani 1984), reflecting the preoccupation with protein in the nutrition literature at the time. Projects aimed to increase the protein in certain crops, such as rice at the International Rice Research Institute (IRRI) (Flinn and Unnevehr 1984) and cassava and beans at the International Center for Tropical Agriculture (Pachico 1984). The International Maize and Wheat Improvement Center was most active in protein-focused research, even recruiting several nutritionists for the project in the late 1970s (Tripp 1984a; Tripp 1984b; Mebrahtu, Pelletier, and Pinstrup-Andersen 1995; Ryan 1984).

These programs on protein-rich crop development did not materialize, however. The nutritionally improved crops tended to create a trade-off between yield and nutrition. For instance, CIMMYT's "quality protein maize" had a substantially lower yield (Mebrahtu, Pelletier, and Pinstrup-Andersen 1995), and protein-rich rice developed by IRRI similarly suffered from the trade-off between grain yield and protein content (Flinn and Unnevehr 1984). As a result of this productivity glitch, most of the protein projects were eventually abandoned (Tripp 1984a; Ryan 1984).

These breeding for nutrition programs also showed the difficulty of cross-disciplinary collaboration. As the agricultural community started to set their goal on nutrition, it turned out that the nutrition community could not agree on the nutritional needs of the Third World. More specifically, as the agricultural community started to take up protein as the key goal, there was growing criticism from within the field of nutrition about the protein fiasco. It was a frustrating time for the agricultural research community. Agricultural experts thought that they had been given a clear mandate from nutritional science to increase the protein content of crops, but the changing focus put agricultural scientists in disarray (Mebrahtu, Pelletier, and Pinstrup-Andersen 1995, 228). One scholar from the CGIAR summarized their frustration:

> At that time protein as the main nutrient that was deficient was at the height of its scientific popularity....It was the protein problem that made the centers take the next step, and research on protein content, the limiting amino acids, and some biological testing was initiated. More than discouraging results in protein quantity and quality, nutrition opinion and criticism induced the next change: the object of increasing production was to increase the intake of energy, now the favorite dietary component. Results showed levels of protein intake to be in most instances adequate or above the needed levels, and the concept of protein quality was somehow lost. (Bressani 1984, 257)

Suffering from the compromised productivity and frustrated by the moving target of "nutrition needs," the nutrition-oriented breeding programs came to be widely acknowledged as a failure in the agricultural research community by the mid-1980s (Omawale 1984; Ryan 1984; Tripp 1990). The idea of breeding for enhanced nutrient content became stigmatized, and the agricultural research community moved back to the traditional productivist research agenda. They reasoned that if people could get large enough crops, their incomes would grow, and then they would be able to buy nutritious food (Ryan 1984, 219). As Per Pinstrup-Andersen has summarized, agricultural researchers at the CGIAR "concentrated on increasing crop yields, ensuring yield stability, reducing costs of production, and protecting the environment, while recognizing that nutritional benefits may accrue *indirectly* from increased crop production and so lower food prices" (Pinstrup-Andersen 2000, 352; my emphasis). The international agricultural community thought that their contribution to improved nutrition should be achieved primarily through higher agricultural yields.

Hampered by the trade-off between yield and nutrition and the difficulty of translating nutritional needs into a practical research agenda, the earlier breeding for nutrition efforts turned out to be short-lived. The nutrition goal was not seen as appropriate for the international agricultural research institutions, except for the general contribution to income growth of developing countries.

Linking Agriculture and Health

In the 1990s, the idea of breeding for nutrition resurfaced. USAID asked the International Food Policy Research Institute to explore possible contributions of agriculture to human nutrition. Subsequently, the IFPRI's Howarth Bouis became the central figure in the promotion of the biofortification concept. But the biofortification concept did not see an immediate acceptance. Given its history, biofortification initially seemed like a bad idea: a new incarnation of an old, failed dream. The IFPRI staff recalled that one of the challenges for biofortification was that the CGIAR scientists tended to assume the inevitability of a trade-off between nutritional value and plant yield from their previous experiences (Bouis 1994). Hence, the upward mobility of the emergent concept of biofortification critically hinged on the prospect of success of the technology. Biofortification promoters needed to provide a compelling argument to the CGIAR that biofortification did not necessarily compromise yields in order to enroll them in its network (Bouis, Graham, and Welch 1999).

Pressured to demonstrate the feasibility of biofortification, the IFPRI looked for research centers that had been working on similar projects. It turned to the Plant, Soil, and Nutrition Laboratory run by the USDA's Agricultural Research Service (ARS) at Cornell University, and the Waite Agricultural Research Institute at the University of Adelaide in Australia. The Plant, Soil, and Nutrition Laboratory had been examining the linkages between minerals in soils and plant nutrients. Waite had conducted studies to improve plant nutrition by breeding for crops with improved mineral uptake from soils (Bouis 1994). The IFPRI hoped that the CGIAR's lack of enthusiasm for nutrition-related projects could be changed by these institutions, which could show some theoretical feasibility of biofortification (Bouis, Graham, and Welch 1999).

The IFPRI organized workshops bringing the CGIAR together with the Plant, Soil, and Nutrition Laboratory and Waite researchers. Researchers from these laboratories presented arguments that biofortification did not have to compromise breeders' traditional goals of yield increase so that they would not be repeating the same mistake they had made with protein-rich crops. They also pointed out the possibility that increasing the micronutrient stores in seeds might increase seedling vigor and viability, improving the performance of seedlings particularly in micronutrient-poor soils (Welch 2002). These presentations from the Plant, Soil, and Nutrition Laboratory and Waite were instrumental in changing the CGIAR's persistent pessimism about nutrient–rich plants. Biofortification promoters reported that the "attitudes towards the micronutrient-dense-seed plant breeding strategy among a core group of the CGIAR plant breeders changed dramatically" after these presentations (Bouis, Graham, and Welch 1999, 6).

Subsequently, the CGIAR embarked on an initiative to develop a five-year plan on biofortification, and at least three CGIAR centers (IRRI, CIMMYT, and the International Center for Tropical Agriculture) signed on (Bouis, Graham, and Welch 1999).

But the scientific persuasion was not the only factor that enhanced the perceived feasibility of biofortification. It was significantly improved by its association with Golden Rice. Indeed, it's now easy to think of it as a product of biofortification research. Currently, Golden Rice is managed by HarvestPlus, whose mission is biofortification, and biofortification researchers often mention Golden Rice as an example of biofortification. Serving as a proof of principle for the biofortification concept, Golden Rice has critically contributed to reducing doubts about the achievability of biofortification.

Yet the widespread association between Golden Rice and biofortification betrays the reality that they started as different programs. Golden Rice research had already started before the concept of biofortification was born. In addition, while the biofortification network was centered on the IFPRI and CGIAR, Golden Rice research had European roots: the project was headed by two researchers: Ingo Potrykus from the Swiss Federal Institute of Technology and Peter Beyer from the University of Freiburg.[6] Furthermore, original motivations were different. The biofortification network grew out of the mandate to shift agricultural research's focus to human nutrition. As such, it could use either conventional breeding techniques or biotechnology, as long as it contributed to improvement in human nutrition. In contrast, Golden Rice was started explicitly as a biotechnology project. One of the funders for the Golden Rice research was the Rockefeller Foundation, whose chief concern was biotechnology's slow adoption by developing countries. Continuing its legacy as the chief architect of the Green Revolution, the Rockefeller Foundation promoted the notion of the "Doubly Green Revolution."[7] They saw that the developing countries were lagging behind in their participation in the "gene revolution," which could be another and perhaps better (ecological, hence "doubly green") agricultural revolution in the global South. It was in this context that they saw Golden Rice as a way to build social acceptance for agricultural biotechnology in the Third World. Gary Toenniessen, who was the director of the food security program at Rockefeller, made this conceptual underpinning clear in 2001:

> [Golden Rice] did not start within a programme that was designed to solve Vitamin A deficiency. Beta-carotene enhanced rice goes back to the beginning of the Foundation's rice biotechnology programme, when the objectives were different. Back then, the Foundation was concerned that new biotechnologies that could contribute to crop

genetic improvement were not being applied to any of the crops or any of the traits that were important in developing countries. In both the private sector and the public sector in industrialized countries, these technologies were being employed for crops important in their own countries that were financially more attractive. Therefore, we started a programme to build capacity in Asia to use these new biotechnologies. (Lehmann 2001)

In this view, Golden Rice was a beachhead that could be used to educate the global South about the merits and potentials of biotechnology. It was in this envisioned trajectory from Green Revolution to gene revolution that Golden Rice was to be utilized.

Despite these different origins, the Golden Rice network and the biofortification network eventually became connected. Golden Rice and biofortification institutionally merged when the management of Golden Rice was transferred from the Golden Rice Humanitarian Board to HarvestPlus in 2001 (International Center for Tropical Agriculture and IFPRI 2002). The task of conducting research for actual dissemination and promotion of Golden Rice in developing countries was now the responsibility of this new organization (IRRI, Rockefeller Foundation, and Syngenta 2001; Schnapp and Schiermeier 2001).

In addition to evidence of feasibility, the excitement about biofortification needs to be situated in the broader trend in international development. One important cultural aspect of biofortification is its hybrid nature, in which the agricultural and the nutritional are conceptually combined. In the 1990s, funding for international agriculture research had dwindled considerably. This was a dramatic change from the 1970s, when the Green Revolution was at its height and agriculture funding had increased by more than 14 percent. But between 1985 and 1996, it grew by less than 1 percent per year (Pardey, Alston, and Smith 1997). The CGIAR system had been particularly hard hit, resulting in significant budget cuts and staff lay-offs (Bagla 1998).[8]

In contrast to the shrinking resources for international agricultural research, the health sector had started to enjoy increased funding from international donors (Okie 2006, 1085). For instance, agriculture used to be one of the largest US foreign aid sectors, but it had been surpassed by the global health sector, whose budget had nearly doubled since 2001 (Tarnoff and Nowels 2004). The World Bank's new commitment in the health sector further accelerated its increasing prominence. The world's wealthiest philanthropic organization, the Bill and Melinda Gates Foundation, also made global health its primary focus.[9] The concept of biofortification was hence an exciting combination of agriculture and health.

Helped by these factors, biofortification succeeded in increasing its public profile and gaining institutional support. In 2002, the CGIAR embarked on a $90 million project called the Global Challenge Program on Biofortification, which aimed to breed micronutrient-dense rice, maize, wheat, beans, cassava, and sweet potatoes with iron, zinc, and vitamin A (Graham 2003). In 2004, HarvestPlus was established to promote biofortification with funding from the World Bank, USAID, DANIDA, the ADB, and the Bill and Melinda Gates. It is headed by Howarth Bouis, who retains his affiliation with the IFPRI and collaborates with the CGIAR institutions, such as the International Center for Tropical Agriculture and IRRI.[10]

By networking with other institutions and adopting Golden Rice as the proof of workability of the concept, biofortification has raised its profile and trustworthiness. In addition, its hybridity in addressing both agriculture and health issues has been useful in increasing its prominence in international development. Global health concerns enjoyed a significant increase in resources in the 1990s, which has been described by one observer as the "golden age of global health" (Okie 2006, 1085). With a renewed emphasis on agriculture in international development since the mid-2000s, biofortification is well positioned to take advantage of its hybrid nature.[11]

Linking the Global North to the Global South: GMO Politics

Besides the agricultural research community, biofortification also has had another set of cheerleaders. For biotechnology promoters who were trying to overcome resistance and opposition to genetically modified organisms (GMOs), the promise of biofortification and Golden Rice provided a useful rhetorical tool to claim wider spatial benefits and moral virtue for biotechnology. Biofortification and Golden Rice embody *benevolent biotechnology*—biotechnology that benefits people in the underdeveloped world by helping them to produce more food and more nutritious food. The discourse of benevolent biotechnology asserts that biotechnology's benefits need to be experienced globally, beyond the modern capital-intensive farms of the developed nations. Indeed, an increasingly prevalent claim made by biotechnology proponents has been that GM crops fulfill the needs of the poor in developing countries. As seen in Monsanto's public relations website, Biotech Knowledge Center, which argues that "biotechnology matters" because "we *can feed the world* for centuries to come and improve the quality of life for people worldwide" (Monsanto Biotechnology Knowledge Center 2001; my emphasis), proponents have argued that GM crops will increase food production and reduce global hunger. In an effort to legitimize such a claim, they have fiercely lobbied at the 2002 World Food Summit to add biotechnology as an option to the resolution on ending world hunger (Carroll 2002). The possible *nutritional* contribution of GM crops has also been used to reinforce arguments for biotechnology. As evident in the following claim made by the Biotechnology Industry Organization, which represented life sciences companies, biotechnology promoters have portrayed GM crops as having a mission to prevent malnutrition as well: "Agricultural biotechnology must be more seriously considered as a significant part of any program to address the *nutritional needs* of the developing world" (Feldbaum 2002; my emphasis).

Marked as a nutritional GM crop for the poor, Golden Rice perfectly fits with the gospel of benevolent biotechnology. With its beta-carotene content enhanced by the inserted daffodil gene, Golden Rice is tasked to tackle vitamin A deficiency in developing countries. Golden Rice's principal researcher and perhaps most vocal advocate, Ingo Potrykus, has painted a dramatic picture of the benevolent potential of Golden Rice in feeding the poor:

> As soon as a novel bio-fortified variety is deregulated and can be handed out to the farmer, the system demonstrates its unique potential, because from this point on, the technology is built into each and every seed and does not require any additional investment, for an unlimited period of time. Just consider the potential of a single Golden Rice seed: Put into soil it will grow to a plant which produces, at least, 1,000 seeds; a repetition will yield at least 1,000,000 seeds; next generation produces already 1,000,000,000 seeds and in the fourth generation we arrive at 1,000,000,000,000 seeds. These are 20,000 metric tons of rice and it takes only two years to produce them. From these 20,000 tons of rice 100,000 poor can survive for one year, and if they use Golden Rice they have an automatic vitamin A supplementation reducing their vitamin A-malnutrition, and this protection is cost-free and sustainable. All a farmer needs to benefit from the technology is one seed! (Potrykus 2004)

Echoing the tenets of benevolent biotechnology, Potrykus paints a utopian picture in which the desperate needs of the poor in the underdeveloped world are fulfilled by Golden Rice. Multiplying at little cost, Golden Rice seed is to deliver its promise of benevolent biotechnology to the poor.

The claim of benevolent biotechnology has also been used by proponents to imply that opposition to biotechnology is tantamount to an ideology that opposes the global South. Readers might recall incidents in 2002 when European nations, which sent African nations money but not GM grain, were accused of "killing people in Africa" because several African nations refused US food aid on the ground that it might contain genetically modified grains. President Bush said "European governments should join—not hinder—the great cause of ending hunger in Africa" (quoted in Clapp 2005, 474). Biotechnology promoters used the occasion to repeat a theme of benevolent biotechnology in which stricter biotechnology regulations were portrayed as harming the poor in developing countries. By asserting that being against biotechnology is being against the global South, pro-GMO groups placed themselves as morally superior to GMO skeptics.

But why does benevolence matter in contemporary biopolitics? What explains such intense moral investment in a biotechnology product? The biotechnology promoters' enthusiastic support for Golden Rice and biofortification as benevolent biotechnology must be understood within the context of a globalized struggle over the diffusion of GM crops in general. The conventional wisdom has been that the struggle over GM crops is principally a matter of a social divide between the United States and the European Union and a trade disagreement between the US state and the EU states. Social differences in US and European attitudes toward GM crops are well documented (Gaskell 2000), and GM crops and foods have been considered a major "irritant" in the EU-US trade relationship since the late 1990s.

After failing to force the European Union and other developed nations to accept GM crops, biotechnology proponents changed their tactics to cultivate markets in the global South (Stone 2002). The major proponents of GM crop products and technologies are placing pressure on developing countries to structure their regulatory systems to adopt the US "substantial equivalence" system, rather than the EU's "precautionary principle" system (Buttel 2003; Schurman and Kelso 2003). At the core of their lobbying is the realization that whether GM crops successfully diffuse globally depends on the Third World (Buttel 2003).[12]

Since the late 1990s, campaigns to promote GMOs to the developing countries have intensified. A good example is Monsanto, whose 2000 annual report made the global broadening of the GMO market one of six critical objectives for the company, singling out several developing nations as targets (Monsanto 2001). It has started to market GM crops in developing countries, notably Bt cotton to India and Indonesia. It is in this context that the actual benefits of GM crops to developing countries has become a fiercely contested issue (Stone 2002; Brooks 2005), and the claim of benevolent biotechnology has become salient for the pro-GMO groups. It comes down to a battle for the moral high ground over who cares more about the poor in the global South.

However, such a claim goes counter to the history of agricultural biotechnology, which has involved research in developed nations and transnational life sciences corporations, and marketing primarily to farmers in developed nations. By far, the biggest beneficiaries of GM crops are North American farmers who produce commodity crops such as soybeans, corn, cotton, and canola. About 45 percent of the total acreage in GM crops is in the United States (ISAAA 2010). Although some "developing countries," notably Brazil and Argentina, have adopted GM crops, they are taken up mainly by large-scale commodity farms, not peasants who practice subsistence agriculture (GRAIN 2009; Binimelis, Pengue, and Monterroso 2009).[13] Most of the GM crops that are planted in developing countries are Bt cotton, corn with herbicide traits, and herbicide-tolerant soy for export and animal feed purposes (Wield, Chataway, and Bolo 2010; Binimelis, Pengue, and Monterroso 2009), so they are not feeding the hungry.[14] Therefore, the claim of benevolent biotechnology that conjures up antihunger effects of GMOs is a huge leap from reality. Golden Rice and biofortification conceal this gap by offering tangible evidence for the ability of GM crops to feed the poor in the less-developed world. As such, biofortification and Golden Rice have attracted enthusiastic support from promoters of GM crops.

Overall, then, biofortification's social versatility can be summarized by two conceptual reconciliations. First, as the embodiment of benevolent biotechnology, biofortification purportedly reconciles the North-South gap in the "gene revolution." Connecting the global North and the global South's biofutures, which had previously been imagined separately, biofortification and Golden Rice have been useful in telling a story of hidden yet continuously unfolding benefits of biotechnology to be extended from the rich North to the poor South. Simultaneously, by embodying agriculture's role in improving human nutrition, biofortification bridges two distinct fields in international development. Such hybridity has brought significant support from those in the international agricultural community, as it provides an opportunity for them to cross a traditional disciplinary boundary to participate in international health issues that enjoyed growth in resources and international prominence.

These two *conceptual* innovations mark biofortification's social appeal, attracting diverse actors and institutions to its network and by linking different yet overlapping calculations and dreams. Indeed, it could be argued that the conceptual innovations are the biggest achievements of biofortification to date, as there has not been meaningful consumption of biofortified products by the poor in developing countries. Their usefulness as a discursive tool was the basis for the heightened profile and stature of biofortification and Golden Rice in the international development scene.

Golden Rice's Slow Circulation

For a product of research in the not-so-lay-friendly field of molecular biotechnology, Golden Rice has enjoyed a great deal of publicity and popular interest.

After the first major publication on Golden Rice in *Science*, in 2000 (Ye et al. 2000), it immediately became global news. The *Washington Post*'s January 14 headline read, "Gene-Altered Rice May Help Fight Vitamin A Deficiency Globally" (Gugliotta 2000). *Time* magazine in July 2000 featured on its cover Potrykus, portraying Golden Rice as "grains of hope" that "could save a million kids a year." (Madeleine 2000). In these media portrayals, Golden Rice was a technological miracle, a scientific breakthrough. Despite such wide attention and praise, the actual deployment of the technology has been quite slow. It has been over a decade since Golden Rice was produced, but the widespread adoption in the Third World that was dreamt of by its pundits is nowhere to be seen. What has happened?

It is not that Golden Rice has not had institutional support. The Golden Rice Humanitarian Board established the Humanitarian Golden Rice Network that included sixteen institutions in Bangladesh, China, India, Indonesia, South Africa, the Philippines, Vietnam, and Indonesia (Potrykus 2004).[15] In 2001, Golden Rice research was transferred to IRRI to develop for practical use in developing countries (IRRI, Rockefeller Foundation, and Syngenta 2001; Schnapp and Schiermeier 2001). Yet the development of Golden Rice has taken much longer than anticipated. The original plan at IRRI was to complete field tests by mid-2003, complete food safety tests by 2006 (Chong 2003), and to have the rice available for commercial release in 2007 (Zimmermann and Qaim 2004). A report in 2003 predicted that plants suitable for field trials should be available within a few years (Rice Today 2003). However, Golden Rice development has lagged behind this original schedule (Paarlberg 2003), and no major developments have been reported to date. More recently, field trials were conducted in the Philippines, but many local groups are questioning its safety (Manila Times 2011).

This has set off a blame game. Golden Rice pundits prefer to argue that some technical glitches and bureaucratic obstacles are holding back the progress of a life-saving technology. Some suggest that a main bottleneck has been the difficulty of breeding with more popular varieties of rice. The original Golden Rice was Taipei 309, because this variety responds well to tissue culture. Yet Taipei 309 is a *japonica* rice, which is not the prevalent rice in Asian countries.[16] Another challenge for Golden Rice is meeting agronomic performance standards in addition to nutritional improvement. Researchers could come up with a wonderful Golden Rice, but what if farmers did not want it? A survey of farmer leaders in the Philippines showed that economic concerns were the most important criteria in decision making on Golden Rice (Chong 2003). Golden Rice has to meet both nutritional and agronomic expectations. The most salient assertion by Golden Rice pundits is that Golden Rice is suffering from excessive regulatory cautiousness. Central to this narrative is the anti-GMO movement that has made governments take an overly restrictive approach to biotechnology. Potrykus's comment below summarizes such a view.

> Considering that Golden Rice could substantially reduce blindness (500,000 per year) and death (2–3 million per year) 20 years are a very long time period, and I do not think that anyone should complain that this was "too fast"! If it were possible to shorten the time from science to the deregulated product, we could prevent blindness for hundreds of thousands of children! However, the next 5 years will have to be spent on the required "bio-safety assessments" to guarantee that there is no putative harm from Golden Rice for the environment and the consumer. Nothing speaks against a cautious approach, but present regulatory praxis follows an extreme interpretation of the "precautionary principle" with the understanding that not even the slightest hypothetical risks can be accepted or left untested, and at the same time all putative benefits are totally ignored.... Golden Rice could prevent blindness and death of hundreds of thousands of children but cannot do so, so far, because risk assessment notoriously is ignoring a risk-benefit analysis! (Potrykus 2004)

Although perhaps unusual in its directness, Potrykus's comment epitomizes proponents' assertion that Golden Rice's surprisingly slow development is due to the anti-GMO movement and resultant biotechnology regulations that are too rigid. Citing the "extreme interpretation" of the precautionary approach in biotechnology regulation that "ignores all the benefits," Potrykus blames regulators for Golden Rice's failure to reach the poor. Furthermore, he echoes the discourse of benevolent biotechnology in charging that a phobia of GMOs prohibits Third World people from receiving Golden Rice's benefits.

Departing from the above view that faults technical and social obstacles, however, we might scrutinize the original claim of Golden Rice's "success" in fulfilling the needs of the Third World poor. Going back to the Ye et al. article in *Science* that first announced the birth of the miracle rice to the world, it becomes clear what Golden Rice actually accomplished was to solve technical riddles that constitute only a small part in the chain leading to vitamin A sufficiency in the Third World.

It is instructive to step back from the assumption of its being a miracle seed and examine the original Golden Rice research to see how it was designed and operationalized. The original research began with the finding that although rice kernels do not have beta-carotene, they make a precursor to beta-carotene (geranylgeranyl diphosphate or GGPP). The research then was narrowed down to find a way to turn GGPP into beta-carotene using enzymes from daffodils. The research question became two technical problems. The first was how to get a daffodil gene into rice. In the early stage of the research, Peter Burkhardt in Potrykus's laboratory tried to use a gene gun, which is a standard way of introducing new genes into plants. But four genes for new enzymes did not get into plants properly. Xudong Ye succeeded with a different strategy by using the plant-infecting bacteria *Agrobacterium tumefaciens* as a vector (Gura 1999).

The second technical problem was which gene construct to use to transform GGPP to beta-carotene. The pathway from GGPP to beta-carotene is as follows: GGPP → phytoene → zeta-carotene → lecopene → beta-carotene. Each conversion needs a specific enzyme. The process of GGPP → phytoene needs phytoene synthase; phytoene → zeta-carotene needs phytoene desaturase; zeta-carotene lecopene alpha needs zeta-carotene desaturase; and the final lecopene → beta-carotene needs lycopene beta-cyclase. Therefore, the process needs four enzymes: phytoene synthase, phytoene desaturase, zeta-carotene desaturase, and lycopene beta-cyclase. The researchers reported in the *Science* article that the "best" line produced 1.6 mcg carotenoids/g of dry weight of rice (Ye et al. 2000). This was really what was reported as the miracle rice that could "save a million kids a year."

Notice the layers of translation from these combinations of enzymes to producing a "miracle rice." The Third World food problem had to be equated to micronutrient deficiencies, and in this case vitamin A deficiency was singled out. Vitamin A deficiency was then translated into the lack of vitamin A in diets, and then further translated into the need for beta-carotene, a precursor to vitamin A. The need for beta-carotene was then translated into a need for a gene that helps the conversion of a precursor to beta-carotene. It was in this successive operationalization that a Third World food need was to be successfully met by a gene from a daffodil.

Once outside the circumscribed definition of food "problem" and "solution," however, such a claim of success seems exaggerated, and even nutritionists have not been entirely convinced of Golden Rice's efficacy. Some have pointed out that carotene, which is the core of Golden Rice's "success," is only the beginning of the solution and is far from sufficient to improve an individual's vitamin A status. The first uncertainty is how much beta-carotene can be obtained from the carotene that is present in Golden Rice. In the original study, the best line produced 1.6 mcg carotenoids/g of dry weight of rice (Ye et al. 2000). However, only about 50 percent of this carotenoid was found to be beta-carotene. This was a disappointingly small amount.

Aside from beta-carotene content, there is the question of bioavailability. Bioavailability is defined as the amount of a nutrient that is potentially available for absorption from a meal and, once absorbed, utilizable for metabolic processes in the body. This is a question of how much beta-carotene from Golden Rice can be actually used by the human body. It has to be noted that beta-carotene is not the same as vitamin A. It is a precursor to vitamin A—meaning that it has to be split by an enzyme to become two molecules of vitamin A. Moreover, not all beta-carotene is utilized by the body. This issue of bioavailability is difficult to determine precisely, not only because it depends on each food, but also because it is not simply determined by total plant mass. Golden Rice needs to be ingested, absorbed, and utilized by the body, and each process is influenced by other meal components, processing, preparations, and even the individual person's biochemical and metabolic characteristics. Particularly problematic for carotene is its need for fat for digestion, absorption, and transport, because beta-carotene is fat soluble. When a diet is low in fat, there is an even bigger obstacle to bioavailability (Gillespie and Mason 1994). Michael Krawinkel, director of the Institute for Nutritional Science at the University of Giessen in Germany, argues that "in countries where the consumption of dietary fat is low, Golden Rice is unlikely to benefit health" (quoted in Schnapp and Schiermeier 2001, 503). Furthermore, New York University nutritionist Marion Nestle says that "many children exhibiting symptoms of vitamin A deficiency, however, suffer from generalized protein-energy malnutrition and intestinal infections that interfere with the absorption of β-carotene or its conversion to vitamin A" (2001, 289). Ultimately, the determination of bioavailability must be made through feeding trials in micronutrient deficient populations under natural living conditions. This complicates the claim of success by necessitating consideration of health conditions as well as different food cultures across and within a nation.

Golden Rice promoters portray the slow deployment of Golden Rice as a question in need of answer. Furthermore, they tend to attribute the gap between the utopian vision and the disappointing reality to a misdirected anti-GMO movement. However, their assumption of Golden Rice's success itself needs to be questioned. Only within nutritionism's narrow purview could the claim of it being a miracle rice make sense. Nutritionism has played a critical role in making the claim of success credible and possible. By focusing on the quantifiable nutrients as the most important aspect of food and the food problem, it has reduced the understanding of the needs of the poor to a particular nutrient. The triumphant narrative of the life-saving miracle rice is what has caught global attention, but the very basis of such a claim—nutritionism—was obfuscated, making it difficult to understand its slow circulation.

Out of Sync in Indonesia

It is perhaps apt here to go back to Indonesia to see how Golden Rice fared there, as Indonesia seems like an ideal beneficiary for Golden Rice. Much of its population depends on rice as a staple food, and awareness of vitamin A deficiency is relatively high among policymakers and even the general public as a result of long-standing educational campaigns and vitamin A capsule dissemination programs by the government. One might imagine that the highly celebrated Golden Rice would be eagerly awaited in a country like Indonesia.

Biofortification promoters surely did not miss Indonesia as a major potential beneficiary of Golden Rice, and they looked to it as a potential ally. The Golden Rice Humanitarian Board and the Golden Rice Network invited the Indonesian Agency for Agricultural Research and Development to be a partner, hoping that they would clear the path for Golden Rice in the country. Marketing efforts targeted at Indonesian researchers were also abundant. There were seminars sponsored by Golden Rice pundits, such as "Biofortification: Breeding for Micronutrient-Dense Rice to Complement Other Strategies for Reducing Malnutrition," which took place at the Ministry of Agriculture in June 2002, with Howarth Bouis from the IFPRI as the keynote speaker. He was again able to promote biofortification in 2004, when he spoke at the prestigious National Workshop on Food and Nutrition (Widyakarya Nasional Pangan dan Gizi).

From the general public's point of view, too, Golden Rice seems to be well-positioned to benefit from positive attitudes to nutrition-targeted applications. One available survey on Indonesian perceptions toward agricultural biotechnology conducted by the International Service for the Acquisition of Agri-biotech Applications found that agricultural biotechnology that is targeted at improving nutrition contents and other food qualities tends to receive a favorable response compared to other kinds of GM crops, and this preference is shared by diverse people, from farmer leaders to general consumers (ISAAA and University of Illinois 2002) (table 7.1).[17]

Nevertheless, Indonesian attitudes toward Golden Rice have been far from enthusiastic for complex reasons. Rather than simply the result of technical glitches or regulatory complications, as the common narrative might suggest, the Indonesian case points to an interconnected web of political and cultural

TABLE 7.1   Indonesian public's perception of different biotechnology applications, % respondents who said each application was "useful"

	MAKE MORE NUTRITIOUS AND QUALITY FOOD	PEST AND DISEASES OF CROPS	PRODUCTION OF MEDICINE OR VACCINES	MODIFYING GENES OF LAB ANIMALS	DETECTING AND TREATING DIS-EASES IN HUMAN
consumers	62	64	70	75	62
businessmen	61	45	14	22	76
extension workers	78	74	22	33	86
farmer leaders	73	53	25	43	69

Source: International Service for the Acquisition of Agri-biotech Applications (ISAAA) and University of Illinois at Urbana–Champaign 2002.

reasons that have limited the appeal of Golden Rice. To start with, Golden Rice became embroiled in the broader contention about genetically modified food in Indonesia. Beginning in early 2000, the Indonesian environmental and consumer-advocacy community began to raise the issue of the safety of GM crops. The first large controversy was with Monsanto's *Bacillus thuringiensis* (Bt) cotton. In 2001, Monsanto received permission for commercial harvest of Bt cotton in South Sulawesi. Environmental NGOs had suspected that Monsanto was planting GM crops through its local subsidiary, PT Monagro Kimia, without formal approval, but the official approval from the government ignited more concerted opposition. They mobilized quite effectively, establishing a coalition of 113 NGOs housed in the Indonesian Consumer Foundation (Yayasan Lembaga Konsumen Indonesia) and in the Coalition for Biosafety and Food Safety, a coalition of 72 NGOs (Jakarta Post 2001d). The NGO coalitions sued the government for issuing the approval without a proper environmental impact assessment. The situation became favorable to the NGOs when news reports started to surface that the harvest of Monsanto's Bt cotton failed although it had been marketed as having a spectacular yield. There was also a report of gene contamination in 2002 (Jakarta Post 2002), although this was not possible according to Monsanto. In addition, Monsanto was found guilty of bribing government officials in relation to the Bt cotton trial in Indonesia (Third World Network Malaysia 2005), to which Monsanto admitted wrongdoing and paid a $1.5 million fine (BBC News 2005). This series of events, particularly the corruption, tarnished the image of Monsanto as well as its cherished technology, genetically modified crops.

Golden Rice is not only constrained by the heated politics around genetically modified crops. While we might expect life sciences companies to be at the forefront of the public relations campaign in the Indonesian market, they have *not* promoted Golden Rice fervently. This is despite the fact that major global players in the life sciences industry—Monsanto and DuPont, for instance—have had a presence in the Indonesian seed market since the late 1980s. When I interviewed the local management of these companies, they recognized the public relations potential of Golden Rice, but they were not engaged in active promotion or lobbying on its behalf. They were more interested in crops that could produce immediate profits. Life sciences companies were pressured to focus on "profitability rather than penetration," as one of them put it. Golden Rice might help them to penetrate the Indonesian GMO market, but it was not likely to yield immediate profits. Following that logic, hybrid corn and soybeans were prioritized. Indeed, this profitability rather than penetration mentality seems to be the underlying dynamic behind Monsanto's eventual decision to pull out of the Bt cotton business in Indonesia after the initial field tests. Ironically, Golden

Rice is orphaned not only by humanitarian groups but also by its most likely protectors—global life sciences companies—in a situation where it is controversial and is seen as having a negative impact on products with a more immediate economic future. For them, Golden Rice was good only *discursively*—as a symbol to be deployed in their global media strategy and public relations campaign. But as a matter of real business strategy, they see little point in pushing for the *actual* use of Golden Rice in countries like Indonesia.

How about nutritional experts in Indonesia? From Indonesian nutritional experts' point of view, there is a mismatch between what they consider the country's needs and how those needs are conceived by Golden Rice researchers. First of all, nutritionists and nutrition-related NGOs understand the importance of vitamin A deficiency, but there is a sense that Indonesia has successfully controlled it by traditional distribution of vitamin A pills. The government has declared success in the campaign against VAD. Moreover, Golden Rice's ability to enhance vitamin A status depends on many variables, and many Indonesian researchers rightly point out that many things remain uncertain about the benefits of Golden Rice (Herman 2002). They have taken a cautious stand, saying that Golden Rice could be an important addition to micronutrient strategies but that more research is necessary and government regulations must be in place.

What about the Indonesian agricultural research community? In the global scene, agricultural researchers seem to be excited about their new health-driven mandate. Yet that does not describe the situation in Indonesia. If Golden Rice does not fit the priorities of Indonesian nutritional policy, it has the same problem in the field of agricultural biotechnology research. The priority of Indonesian agricultural research is more on abiotic and biotic stresses on plants than on nutrition. This focus on productivity and yield has historically been the case in Indonesia and remains so. For instance, according to Indonesia's Repelita VI, 1993–98, the ultimate goal of agricultural biotechnology research was to achieve and maintain self-sufficiency in food production, develop agroindustry, increase efficiency in using biotic and abiotic resources, and increase crop and animal production. The Agency for Agricultural Research and Development, which is the central biotechnology research institution in the country, has publicized a similar emphasis on productivity. The agency's 1999–2004 plan was for a broad research focus on yield increase and integrated plant management. Specifically for biotechnology research, the agency decided to focus on disease-resistant crops and germplasm conservation (Bidan Litbang Pertanian 2004). The strategic planning for the period of 2005–9 continued with similar priorities of high yield, biotic stress resistance, abiotic stress resistance, and fit with consumer tastes and preferences (69). Biotechnology research priorities for the same period shared similar objectives (75–76). Hence breeding for nutrition, as embodied by Golden Rice, has had little resonance with the overall direction of agricultural and biotechno-logical research in Indonesia.

This is not to say that nutrition-driven agricultural research has been non-existent in Indonesia, but such projects are very few and have tended to come from international donors. For instance, one nutrition-based initiative was breeding for iron-rich rice. As a part of the global program coordinated by the ADB, MI, Danish International Development Agency or DANIDA, USAID, and IRRI, Indonesia was asked to participate along with several other developing countries, and the government agreed to allocate two Indonesian researchers (Hunt 2001b). They published results on screening for germplasm for high iron and zinc contents, the examination of their growth in field conditions, and a nutritional study (Somantri and Indrasari 2002; Somantri and Indrasari 2003; Somantri and Indrasari 2004). Nonetheless, this iron-rich rice project remained outside the mainstream of Indonesian agricultural research and was seen as one of many donor-driven projects.

Furthermore, although the Golden Rice promoters saw a major advantage of Golden Rice in the fact that it was rice—one of the main staple foods in Asia—this very fact seemed to be a sticking point from the Indonesian researchers' point of view. One nutrition expert at an NGO suggested: "Rice is our staple food. So anybody trying to manipulate rice should have strong support from policymak-ers. It will endanger the whole nation. I think people tend to be conservative when it comes to rice. Because it's rice."

Some researchers described rice as "a political commodity" and argue that anyone who tries to meddle with it takes a political risk. Rice was also described as a "way of life," suggesting that consumer resistance might be high. The fear of messing with rice is perhaps mysterious from a perspective limited by nutrition-ism, which sees rice (or any crop, for that matter) as a mere "vehicle" for nutri-ents. Yet if one thinks of food as a cultural as well as a nutritional entity, rice has perhaps one of the most tangled and complicated sets of meanings of any food in Indonesia. Rice features prominently in mythologies about the goddess Dewi Sri in many parts of Java, where rice is said to have sprouted from the dead body of the goddess, and it is seen as a gift from heaven or the underworld (Wessing 1990; Heringa 1997). In Bali, too, rice is seen as a gift from the gods, and rice production is a practice that needs not only human but also gods' hands (Howe 1991). According to anthropologists, rice is also intricately linked with ethnic and sexual identities (Colfer 1991).

Rice is also infused with political tension for the elites of the society. With its symbolic and cultural significance, throughout history rice has been a poignant medium for people to express their anger with a ruling regime such as by rice riots.[18] Given the rich symbolic value of rice in Indonesia, it is not a surprise that rice is deemed highly political and sensitive. Indeed, not long ago, Indonesia witnessed the potent symbolic power of rice in shaping politics when the Sukarno government's fall was closely tied to its failure to contain the skyrocketing rice price. That rice is a political commodity is well understood by Indonesian elites. Rice and political and regime stability are intricately linked.

Biotechnology proponents and news media have portrayed Golden Rice as achieving something grand—fulfilling the needs of the Third World and saving malnourished children. Golden Rice was used to showcase the broader utility of biotechnology for the Third World, as a technology to provide more nutritious food. Within this triumphant narrative, Golden Rice's slow circulation comes as a surprise and has been attributed to some unfortunate technical difficulties and misdirected social skepticism about biotechnology. Yet once one looks beyond nutritionism, this heroic biotechnology was too far removed from those it intended to impress with its benevolence. Golden Rice pundits' moral claims regarding the well-being of Indonesians seemed disingenuous to civil society groups, who saw the rice as a beachhead for GMO and its politics. In this case, Golden Rice was also out of sync with business and scientific interests.

In a discussion of the "microscopic view" of experts that has caused many modernist state projects to fail, Scott (1998) suggests looking for what "fell out" of that way of viewing, in order to understand the peculiarity of this vision. As in many other cases of nutritional fixes, the Golden Rice story shows how much has been rendered invisible by nutritionism. In the experts' dialogue and the celebratory remarks by biotechnology supporters of Golden Rice, food is implicitly medicalized and considered a mere amalgamation of nutrients. Of course, that Golden Rice was rice did matter, but only to the extent that it promised that a certain amount of the nutrient would be carried to the target population. What "falls out" from nutritionism's portrayal of food as a mere vehicle of nutrients is an understanding of food as a deeply cultural and politicized commodity. In ignoring the complicated layers of meaning of rice, Golden Rice stands awkwardly as a well-intentioned, yet inappropriate, "gift" from the international community.

Moral Politics of a Nutritional Fix

Golden Rice shares familiar themes with the stories of the quest for magical solutions that we have seen here. It is one of the latest examples in the long history of nutritional fixes. As such, the story of Golden Rice and biofortification provides an illuminating case to further explore the structure of nutritionism and nutritional fixes. Particularly, I want to pose the question of why nutritional fixes are so persistent, dodging and eluding opposition. If so much "falls out" from the circumscribed view enabled by nutritionism, what explains the persistence of nutritionism and nutritional fixes? Of course, there are various reasons why a nutritional fix is powerful. It is, first of all, elegant in its simple statements and appealing in its calculability. It is embedded in a broader scientific reductionism with a long history. It is also easy to see why bureaucratic institutions might find it helpful to have the focused representation of food problems afforded by nutritionism because it facilitates policy planning and implementation. Nutritional science, as we have seen, also has a disciplinary stake in being able to sell nutrition in the form of nutritional fixes. But this chapter has also highlighted another critical factor in nutritionism's tenacity: its moral claims.

Indeed, the intensity of moralistic claims around Golden Rice is a curious feature when compared to the standard talk about GMOs, which tends to be framed in terms of the necessity of basing decisions on science and rationality. When GMOs started to become a socially controversial issue, governments, scientific bodies, and the industry called for a "rational" discussion, and "risk assessment" was the language of choice, in which the benefits and costs of the technology were to be gauged in an ostensibly scientific manner.

The profusion of moral claims surrounding Golden Rice and biofortification attests to nutritional fixes often conjuring up a nutritional utopia where the world food problem is solved through amazing modern technology. An important actor in this grand narrative of nutritional utopia is the hungry to be fed, victims to be saved by nutritional fixes. As food historian Warren Belasco points out, the West's imaginaries of a cornucopian utopia is often accompanied by the representation of the developing world as a dystopia (Belasco 2006, 116, 168).[19] Ingrained in the West's historical understanding of the world food problem is the notion of "the poor" of the developing world as the hungry to be rescued and emancipated by the West, and it is for these imagined beneficiaries (and in honor of the West's benevolence) that nutritional fixes are celebrated. Continuing this historical pattern, victimization of the Third World poor is conspicuously present in the imaginary of biofortification/Golden Rice. Recall Potrykus's assertion that "a hundred thousand poor" in the developing world were to benefit from a "single" Golden Rice seed. In this comment, the conceptualized poor in the Third World underlines the moral claim of those promoting the product, as they are seen as waiting to be rescued by the amazing technology.

The nutritional utopia creates a situation in which criticism of the narrowness of nutritional fixes can be portrayed as an attack on the benevolent *intent* of the technology. This results in a decrease in the moral authority of the critics. Golden Rice has plenty of examples of such slighted moral competition. As we have seen, one important political corollary of the discourse of benevolent technology was to make being anti-GM equal to being anti–global South. Skeptics of Golden Rice and biotechnology were subject to strong moral condemnation. In the words of Potrykus, "Those opposing use of the rice in developing countries should be held responsible for the foreseeable unnecessary death and blindness of millions of poor every year" (quoted in Schnapp and Schiermeier 2001, 503). And this attack was not limited to Potrykus, who has arguably been the most fervent crusader for Golden Rice. Where we began this chapter—at the Copenhagen Consensus Conference—economists deployed Golden Rice as a way to malign European trade policies on GM crops. In a report prepared for the Copenhagen Consensus Conference, Kim Anderson and L. Alan Winters (2008, 33) wrote, "This new technology [Golden Rice] has yet to be adopted, however, because the European Union and some other countries will not import food from countries that may contain GM crops even though there is no evidence that GM foods are a danger to human health. The cost of that trade barrier to developing countries has been very considerable." In this discourse, skeptics of a nutritional fix are worse than simply being against science and technology—they are also constructed as costing the lives and well-being of the poor in the developing countries.

Such claims of moral righteousness are enabled by the purported success of nutritional fixes to "solve" the food problem. Supporters of nutritional fixes assert that they offer a practical solution to hunger and malnutrition. By claiming that nutritional fixes actually get things done, supporters often gain high moral standing, and so they become hard to fault. In this process, the mechanism that enabled the claim of success—nutritionism—becomes hidden. This was evident in the case of Golden Rice. Its research was really about the conversion of a precursor to beta-carotene to beta-carotene, and Golden Rice had many limitations as to its effectiveness in tackling vitamin A deficiency. Hence Golden Rice's claim of success was founded on narrowly defined technical problems. Yet once success is claimed, anyone who attempts to scrutinize it becomes an unappealing nay-sayer, an unhelpful detractor. That success in meeting the needs of the poor is partial at best and only justifiable within the narrow purview of nutritionism becomes difficult to convey. Inasmuch as the nutritional trope tells the story of tragic victims to be saved by nutritional fixes, unpacking the underlying nutritionism behind success becomes tricky, since such a move is easily portrayed as against the beneficiary (the poor in the global South) and against morality (the mission to feed the hungry).

As we have seen in the Indonesian response to Golden Rice, however, a nutritional utopia conjured by nutritional fixes is founded on a fragile and limited basis, as it fails to address the cultural, symbolic, and political significance of food. As the story of Indonesia shows, actual responses by the presumed beneficiaries might be ambivalent or even hostile to the imposition of nutritional fixes. Nonetheless, curiously made difficult in food politics is any criticism of nutritionism and nutritional fixes, in part because of the distant, yet emotionally powerful imagery of the poor in the Third World. These poor constitute a part of the powerful mechanics of nutritionism by symbolizing the "victims in waiting." The success of nutritional fixes remains unquestioned as long as these "poor" in the Third World remain abstract, distant inhabitants of the nutritional dystopia.

Conclusion

The attention of the nutrition community and the resources of donors are more attracted by the glamour of micronutrients, a largely technical and often top-down solution, than by the politically sensitive business of poverty alleviation, people's empowerment, and equity.

—C. Schuftan, V. Ramalingaswami, and F. Levinson, *Lancet*, 1998

A mother of four children (eleven, six, three, and two) is talking to me in a Jakarta neighborhood. She tells me: "My husband is a clerk—works at a store. He gets 15,000 Rupiah a day. But we need at least 10,000 Rupiah for food. We eat *nasi uduk* [rice steamed with coconut] or noodles in the morning, *kentang kukus* [steamed potato] for lunch, and, for evening, sometimes we eat, sometimes we don't." Although she looks healthy, and I note she is wearing pink lipstick, the two and three year olds are very thin and seem to have health problems. Her place is a two-story shack and very small, perhaps six by ten feet, but her parents and her three brothers stay there as well. They don't have a bathroom, and MCK (the public bathroom) tends to be full, so the family bathes in the river, which to me looked heavily polluted with greenish-brown water and floating garbage. I wonder about its impact on infections that could be related to micronutrient deficiencies. Besides her husband's meager wage, this mother complains about the cost of schooling. For elementary school, the monthly fee is about 15,000 Rupiah, but she also has to pay for transportation (kids take *mikrolet* bus) and books ("could be 100, 000 Rupiah—they need to be Xeroxed"). She also laments the high cost of rice. She also needs to buy water for cooking and drinking.

This mother, like many other mothers I interviewed, used Promina and Nestlé porridge for her children's meals. She also received the World Food Programme's fortified cookies. Obviously, these products added micronutrients to their diet. Many mothers liked the aid food because, after all, who does not like free stuff? On some level, it is easy to see how such programs have become popular in developing countries. However, such focus on "missing micronutrients" and the use of fortification and biofortification as the solution has important consequences. Like other charismatic nutrients and nutritional fixes that came before them, micronutrients and fortified and biofortified products are often so simple, straightforward, and tangible that other possibilities become invisible or unattractive. By defining missing micronutrients as the problem, a world is created where issues such as poverty slip away from the policy discussion. By channeling resources into the delivery of micronutrients, the opportunity to address other issues as the underlying cause of hunger is lost. By defining poor women as the problem and the object of policy intervention, there is little incentive to include them as agents of policy with insights and valuable inputs. Various issues that the mother spoke of in the above interview—the rising price of food, school fees, low wages, costly drinking water, and the lack of hygienic living—are left unaddressed when fortified cookies and porridges dominate the conversation about food policy. These remain as the unfortunate, but remote, background to hidden hunger.

Nutritionism brings a subtle but profound change in how we talk about food and health, and consequently how food is made a target of particular kinds of interventions; it has thus changed the landscape of food politics in developing countries. I have attempted to show the operation of nutritionism in the fields of international policy (chapters 2 and 3), national policy (chapter 4), and in three commodity examples in Indonesia (chapters 5–7). These chapters highlighted the indispensable role of nutritionism in translating and acting on food insecurity in the developing world. By privileging nutritional science as the foremost authority to diagnose and control the Third World food problem, nutritionism has become a particular lens through which we see food insecurity in developing countries. It is an art of managing the representation of the "problem" of the Third World food and people.

What is striking about nutritionism is its influence in the powerful institutions of society, from the market to government to science. In each field, nutritionism has an important function. I have shown how nutritionism fits well with the logic of the market. That scientific reductionism translates well into economic reductionism is not new in food politics. Agrofood scholars such as Kathleen McAfee (2003) have pointed out the link between biotechnology's reductionist tendency and its business potential, and historians of nutritional science such as Rima Apple (1996) have found that the development of nutritional science in the United States has been linked to the historical growth of the market for vitamin-based products. Scientific reductionism is also integral to the commodification of food (Friedmann 1999; Beardworth and Keil 1997). By erasing the "social life" (Appadurai 1988) of food, reductionism refashions food and agriculture into manipulable and tradable "things" amenable to the logic of the market place (Goodman and Redclift 1991). Furthermore, we have seen how nutritionism particularly resonates with neoliberalism. The food industry is well positioned to argue their expert status in adding nutrients to food products and marketing them, and partnership with the private sector and market-based solutions to social problems have been increasingly seen as preferable in international development.

We have seen that the cultures of government and bureaucratic organizations have further provided a fertile field for nutritionism. Nutritionism, to borrow the title of James Scott's (1998) book, helps one to "see like a state." It makes a complex food problem legible, manageable, and controllable by simplifying it into a matrix of biomedical parameters. For instance, we have seen how bureaucrats, international organizations, and nonprofit organizations have found the advantages of being able to simplify the food problem into a nutritional problem, because it has meant that the problem could be operationalized as a matter of identifiable and quantifiable nutrients. Such "translations" have made designing and evaluating food policy programs more manageable and, importantly, have made the claims of these programs' success and effectiveness more convincing. In international development, furthermore, the motivation to streamline food policy programs also has come from the recent drive for "evidence-based" programs as a part of larger neoliberal accountability politics (Graham 2002). In this context, nutritionism has enabled development program officers to articulate how much nutrient-specific programs, such as WFP cookies and instant noodles for women and children, have delivered to the target population, if not how much the programs have actually improved their health. Governments have thus also been able to obtain "objective" measures of food projects that satisfy the requirements of international donors who worry about so-called development leakage and corruption.

Furthermore, from nutritional science's point of view, nutritionism is tremendously important for academic disciplinary "boundary making" (Gieryn 1983). That is, nutritionism confers the ability to define the problem as a "nutritional" one. Nutrients, such as iron and vitamin A, can be thought of as "actants"—things that help scientists build technoscientific networks—because they embody their unique object of study and social contribution (Latour 1987). By carving out a field of expertise that belongs solely to nutritional scientists, nutritionism has helped to elevate the prestige and relevance of their field. By formulating the food problem as a nutritional problem, nutritional scientists' prestige increased vis-à-vis other academic disciplines and in the international food community. And this is a particularly relevant issue because, as James Levinson (1999) points out, nutritional science has long striven to gain legitimacy and relevance in the business of development. In Indonesia, for instance, the nutritional sector had struggled against the agriculture and population sectors and their neo-Malthusian paradigm. For experts within the development apparatus or in developing countries, their relevance to "development" has critically shaped their professional fates, and nutritionism helps to assert nutritional science's contribution in the international development sector.

In contrast to the converging forces of these powerful institutions in society—the market, the state, and science—what is absent are the poor and the hungry themselves. One important consequence of nutritionism that I want to underscore is that it tends to create a space where only experts can define and prescribe for the Third World food problem. For instance, we saw how various charismatic nutrients in their time dominated food policy debates, foreclosing other possibilities to understanding the food problem. The absence of the poor themselves in telling their stories of hunger and malnutrition is hard to ignore in all the case studies that I examined in the context of Indonesia, although their well-being is discursively highlighted by the expert community and the food industry. Instead, the feeding and dietary practices of the poor and the hungry themselves, particularly women and mothers, come under scientific scrutiny. Women tend to be held accountable for not feeding children and their family properly while the food industry emerges as the savior of the hungry and the malnourished and as a suitable partner in food policymaking.

The power of such exclusive expert discourse derives not only from its institutional base in the market, government, or science, but critically from being taken for granted and naturalized. Yet experts do not have to be the only legitimate voices in defining the nature of the problem and in creating solutions. This book's objective has been in part to describe the phenomenon, but also to point out its particularity and open-endedness. There are moments in which the blind spots of nutritionism surface, revealing the contradictions and tensions within, the destabilizing moments, and hinting at alternative spaces.

Contradictions of Nutritionism

Nutritionism has many blind spots, and I have highlighted their political and social implications. As an adjunct professor at Tulane School of Public Health and a founding member of the People's Health Movement, Claudio Schuftan (1999) notes, "One can rightly wonder if this [micronutrient focus] represents an attempt to avoid the more difficult choices and challenges in the battle against malnutrition and—in the name of nutrition—focusing more on its more achievable areas of impact thus choosing the relatively easier path to staying involved in nutrition work." But even when remaining within the dominant biomedical model, there were many issues that I thought paradoxical, and in this section, I want to summarize four issues.

First, when malnutrition is addressed by focusing on the micronutrient makeup of food, what is often hidden from view is the general lack of food and calories. In fact, some experts have criticized the focus on micronutrients by suggesting that it has unjustifiably shifted crucial resources away from combating protein and calorie malnutrition. In the *Lancet*, Schuftan, Ramalingaswami, and Levinson (1998, 1812) criticized the current trend and pointed out that "it is clear that we have, all too often, neglected the over-riding issue of inadequate calorie intake and its determinants which continue to take such an enormous toll on vulnerable populations." For them, the popularity of micronutrients is rooted in a quick-fix approach, while protein-energy malnutrition, which is more difficult to tackle, has been neglected. We saw this concern materialize in Indonesia. While the Indonesian government was spending much of its nutrition budget on fortified food, and nutritional experts were discussing the need to move to a "new paradigm" of micronutrients, local newspapers exposed various cases of the old type of hunger, with its visible signs of malnutrition (TEMPO 2005a, TEMPO 2005b, GATRA 2005). The lack of attention to the social causes of hunger and food insecurity under nutritionism can even be seen as leading to the persistent vulnerability of marginalized communities to protein-calorie malnutrition.

Second, because nutritionism narrows attention to a "lack" of nutrition as the problematic, the emerging issue of obesity—or overnutrition—is not adequately addressed. Obesity has become a global concern both in developing as well as developed nations, and many warn of the deleterious effects of the "nutrition transition" (Haddad 2003) in the global South. In Indonesia, too, obesity has also been on the rise since the 1990s (Soekirman et al. 2003; Atmarita 2005). With their focus on the issue of deficiency, nutritionism has difficulty dealing with the coexistence of malnutrition and overnutrition in a comprehensive and holistic manner.[1]

Third, we have seen more specific cases of the paradoxical implications of nutritionism. For instance, evident in the case of fortified baby food promotion was how a nutritional fix might solve one problem while creating or exacerbating others. The promotion of fortified baby food might make sense as a micronutrient strategy, but it can undermine the message of breast-feeding promotion. That is highly paradoxical, as breast-feeding promotion is the professed goal of many international organizations and national governments, including Indonesia. Of course, in the minds of nutritional scientists, there is no conflict: an enlightened mother exclusively breast-feeds for six months *and then* adds commercial complementary food that is properly fortified. Therefore, the promotion of commercial fortified baby food and the espousal of breast-feeding do not present any contradiction. However, this clear line between before and after six months of age has little realistic application. The breast-feeding statistics summarized in chapter 7 reveal that many mothers do not breast-feed at all or stop or reduce breast-feeding earlier than the recommended six months of age. When I interviewed mothers in the Jakarta slums, it was evident that the medically correct rule was not well understood. The experts' endorsement of fortified commercial baby food might be self-defeating, as what actually lingers in the consumers' consciousness might be the message that commercial food is more nutritious and optimal.

Similarly, the distinction that experts have made between "properly fortified" and "junk" food was not so self-evident in the eyes of the mothers whom I interviewed. Technically, nutritional science only recommends properly fortified products according to the daily requirement of each age group and the prevalence of micronutrient deficiency in the country. For instance, a lot of thought went into the formulation and amount of nutrients to be added to the World Food Programme's fortified cookies and instant noodles so as to meet the nutrition requirements of Indonesian children. Although science may draw a clear line between products with proper fortification and products without, it is unclear how that distinction plays out in consumers' minds. While consumers are very sensitive to the overall nutrition appeals in advertisements, understanding of nutrition information is actually very limited. Therefore, when I asked interviewees who received the WFP's fortified aid food what they would do after the end of the program, they simply said, "Oh, we can buy at *warungs* [small vendors]" or "It's sold at *warungs*" or "Like Marie and Roma [types of cookies]." That is to say, in the minds of consumers, the WFP cookies that are "scientifically" properly fortified are no different than regular cookies without proper fortification. Officially, nutrition education was to accompany the WFP's food distribution, but it was not frequent (many of my informants did not recall it), and it did not emphasize the difference between fortified food and regular cookies and instant noodles. Regular cookies like Marie and Roma also have confusing claims such as "high in calcium" and "vitamins" on their packages. This kind of food aid, then, can be seen as creating the *habit of eating cookies* rather than the habit of eating properly fortified cookies. This was exactly what manufacturers expected. Such concern about habituation is even more justified when the majority of parents let children decide what kind of snacks to buy, which was suggested in the interviews. Therefore, to expect that this kind of food aid instills the habit of eating properly-fortified food seems far from realistic. This is also alarming based on the fact that a growing number of studies are now finding that children with *malnutrition* already eat a lot of snacks, such as fried chips, cookies, and cold drinks (Sudjasmin et al. 1993).

Fourth, nutritionism's celebration of fortification and the partnership with the private sector might not reflect the reality of the volatile global market and the behavior of private corporations in it. As I have mentioned, the food crisis in 2007–8 significantly increased the prices of commodities that are used for fortification such as wheat, oilseed, and sugar. It is ironic that when people's access to nutritious food was acutely strained, the very food products that were supposed to carry nutrients became too expensive for the poor in many countries. Furthermore, the private sector might not prove a reliable partner in fortification projects, particularly when their primary mission—profits—is jeopardized. In Indonesia, the milling industry betrayed their earlier commitment and lobbied for the suspension of fortification when it started to see it as unfavorable for business. This illustrates how the sustainability of the market-based solution needs to be scrutinized and assessed in a way that considers the increasing volatility of the global food market.

Nutritionism is seductive because it offers a technical and seemingly straightforward framing of the food problem and quick nutritional fixes for it. Because of this and its analytic limits one is blinded to the totality and complexity of the problem. As many agrofood studies scholars have pointed out, food problems in the developing (and developed) world are related to various factors including the structure of global capitalism, the system of economic and political control, and the culture of food marketing and consumption. Furthermore, nutritionism's strong belief in the power of modern science in shaping people's conduct often turns out to be naive. Policies based solely on nutritional calculations ignore the basic facts that people eat and feed for many reasons, including, but not limited to, nutrition and health. Nutritionism might be productive for short-term policy planning, but the long-term consequences of resorting to nutritionism need to be seriously considered. The global food problem is not the simple sum of several nutritional deficiencies.

A Space to Imagine the Alternatives

By dislodging the naturalized correspondence between the "reality" and any problem definition, Foucault's concept of problematization is helpful in imagining other possibilities for defining the problem and solutions. My argument in this book, that nutritionism has constrained the food problem definition and given rise to nutritional fixes, must be seen as a critical intervention to open up a space for imagining an alternative problematization in food politics.

In the introduction, I pointed out how the micronutrient turn in the 1990s did not depart from the productivist paradigm in a profound manner, despite the seeming differences on the surface. As much as it looked like a radical break from productivist policies, the micronutrient turn failed to mount a thorough criticism of the mainstream discourse on food insecurity.

Where then do we find a truly radical praxis, an alternative to the scientized views? We can find a radically different apparatus of the contemporary food problem in grassroots social movements. "Alternative agrofood movements" (Allen 2004) in many countries have created various programs to improve the food system (Allen et al. 2003; Henderson 2000). Many peasants in developing nations are organizing themselves with concepts such as "food sovereignty" (Patel 2007) and the "right to food" (Rocha 2001). Movements to fight corporate control of plant germplasm (Shiva 1997) and land monopoly (Lappé and Lappé 2002) have also been active in many parts of the world.

These grassroots movements' conceptualizations of the food problem and its solutions tend to differ from the reductionist ones described in this book. From the point of view of the grassroots movements, the food problem is not simply a nutrition gap or a productivity gap. Instead, they have critiqued the growing power of transnational agribusinesses, modern agriculture's environmental pollution, and agribusiness's harsh treatment of workers and animals. They have argued for the value of local food and the abolition of the international trade agreements that have assisted agricultural trade liberalization. They have advocated for the rights of small farmers and the importance of their control over land and other productive resources. From their point of view, technical fixes such as the Green Revolution package, fortified food, and Golden Rice, fall far short of addressing the problem of hunger and malnutrition.

I refer to these social movements, not only because they point to the possibility of different solutions, but because they are putting Foucault's problematization concept into real action in their own praxis. While they might not use academic jargon, activists have figured out that the space to define the problem itself has enormous implications. Among various food movements, the most radical conceptual counterpart to nutritionism can be drawn from the "food sovereignty movement."[2] First used by the peasant-based Via Campesina movement in the 1990s, food sovereignty refers to the "right of peoples to define their agricultural and food policy" (Desmarais 2007, 34). The movement has spread globally, and in Indonesia, too, there is a growing movement using the concept, now translated as "*kedaulatan pangan*" or food sovereignty (Winarto 2005). Pointing out that hunger was often used as a justification to push for trade liberalization, agricultural modernization, and privatization, the movement has forcefully asserted the central importance of agriculture and small-scale farmers for combating hunger and malnutrition.

The food sovereignty movement seemingly shares the same goal of the eradication of hunger and malnutrition with mainstream food insecurity discourses. Yet it profoundly differs in its approach. The movement argues that at the core of the world food problem is not the lack of food but the lack of "self-defined ways to seek solutions to local problems" by local communities (Windfuhr and Jonsén 2005, 15; Patel 2007). The Via Campesina's definition of food sovereignty, the "right of peoples to define their agricultural and food policy," encapsulates their insistence that the food problem is about people's self-determination and power, which starts at the level of diagnosis of people's own situation and of problem definition. As Patel (2007) observes, it is "a call for a mass re-politicization of food politics, through a call for people to figure out for themselves what they want the right to food to mean in their communities" (91). These activists have argued that it is the lack of autonomy and participation in defining the problem (and the solution) that ought to be considered the core of the food problem.

With the emphasis on participation and self-determination, the marginalized and the poor are no longer pigeonholed as victims. As we have seen, nutritionism has frequently marked women as victims (remember the concept of *biological victimhood*) and as recipients of food policies. In contrast, the food sovereignty movement has been able to address the importance of female participants in improving the food system rather than their victimhood. Many women have participated in decision making, helping to articulate the movement's goals and to include gender equity as an important aspect of the food sovereignty concept (Desmarais 2004). Listen to the women in the food sovereignty movement, who forcefully declared in 2002: "We women, from various continents, representing countries of the South and the North, demand the right to be free from hunger for every woman, man, and child. We ask for the right to govern our livelihoods, and to have access and maintain control over our lands, waters, seeds, and all resources which are basic to our and our communities needs."[3] Defying nutritionism's characterization of them as passive, biologically determined victims of malnutrition, these "victims" demand "the right to govern our livelihoods."

This issue of participation and representation leads us back to my most profound criticism of the scientized view of food insecurity: its depoliticizing effect. For instance, nutritionism's reductionist, technocratic, and ahistorical tendency is seductive because it can avoid more structural and hence politically sticky issues. Once within the worldview of nutritionism, it is easier to evade a social view of hunger and malnutrition that would necessarily include macroeconomic and political issues of poverty, inequality, and marginality. The poor are advised to eat better—read "more nutritious"—food rather than blaming the government, the world order, and capitalism. Nutritionism's individual level of analysis implies individualized responsibility, too. Locating the cause of micronutrient deficiencies, malnutrition, and hunger with the individual, rather than at the social level, nutritionism tends to shift blame onto people for making bad choices. Recall, for instance, how infant micronutrient malnutrition is typically seen as the mother's mismanagement of feeding practices, rather than the outcome of structural constraints that have limited mothers' feeding choices and living conditions. This is not to say structural factors such as poverty and inequality are not acknowledged. Rather, they are not considered the primary causes of the food problem. Cast as distant factors whose relevance to policy is not immediate, structural factors are often viewed as hindrances to getting things done. The food problem became a problem *of* food, rather than a problem *around* food. In other words, the nutritional makeup, rather than the political economy, of food defined the parameters of the possible conversations.

Paradoxically, we are, then, in critical need of languages to talk about a *food problem beyond food*. Yet this is difficult within a scientized view of food insecurity, as it tends to close, rather than open up, a space for broad-based social participation in food policy talks. For instance, nutritionism simplifies the policymaking process, not only by reducing it to biochemical aspects, but also by reducing the range of actors who are considered relevant. Scientific and technological representation replaces political representation, giving science and policy experts a wide space to represent the food problem, while leaving little room for citizens. Yet as women activists in the food sovereignty movement proclaim, the food problem is about livelihood, including, but not limited to, nutrition and food. It is only when we limit the discussion to the technical aspects of food that food reform becomes the de facto territory of experts.

Experts have triumphantly claimed to have uncovered hidden hunger (micronutrient malnutrition). Yet the growing demand for self-determination and democratic participation from people in food movements powerfully shows what is hidden and marginalized by such scientific triumphalism. The food problem is not only about the lack of science and modern technology, it is about livelihood and sustainability. It is not only a scientific question, it is a political question. We can truly "uncover" hunger and malnutrition, not by the national food balance sheet, dietary surveys, or biochemical experiments, but only by listening to people's—and particularly women's—voices.

Notes

1. UNCOVERING HIDDEN HUNGER

1. The Grameen Bank's founder, Muhammad Yunus, received the Nobel Prize for his work in microfinance. The joint venture is called Grameen Danone Foods.

2. For instance, in the United States, salt iodization began nationwide in the 1920s (Backstrand 2002), and iodized salt accounted for 90–95% of salt sales (UNICEF and Micronutrient Initiative 2003). Vitamin D's link with rickets was discovered in 1924 (Carpenter 2003c), and large-scale milk fortification with vitamin D was soon developed (Bishai and Nalubola 2002). Thiamine (B_1) was synthesized in the 1930s, and thiamine fortification of flour began soon afterward (Bishai and Nalubola 2002). In addition to voluntary fortification schemes, many states started to require flour fortification in the 1940s (Park et al. 2000).

3. The concept of nutritionism has been used by Gyorgy Scrinis (2008) and also popularized by Michael Pollan (2008). Other scholars have discussed the growing power of nutritional science (Belasco 1993; Dixon 2002; Dixon and Banwell 2004; Levenstein 1993). For instance, Jane Dixon and Cathy Banwell use the term "nutritionalization" as "the growing dominance of nutrition and health considerations in all facets of dietary discourse and of the food supply itself" (Dixon and Banwell 2004, 119). The concept of nutritionism denotes a particular tendency influenced by modern nutritional science but does not assume that all nutrition-related concerns have this tendency. The concept also equips us to highlight where such tendencies surface, rather than project a sweeping shift.

4. Of course, nutritionism is not the only reductionist tendency in the agrofood system. Modern agricultural technologies are rooted in a reductionism that disembeds farming from its local ecological and social contexts (Scott 1998), and the contemporary advocacy of genetically modified crops is closely linked to molecular reductionism (McAfee 2003; Sarkar 1998).

5. See, e.g., Ferguson (1990), Mitchell (2002), Agrawal (2005), and Li (2007).

6. Nutritional fixes can be considered a version of the "technological fixes" theorized by physicist Alvin Weinberg in his science and technology classic *Controlling Technology* (1991). Weinberg famously advocated technological solutions for social problems. Rudi Volti (1995) argues that technological fixes have not been able to solve underlying problems and that technology has always been influenced by power relations.

7. For instance, in their review of existing fortification projects around the world, Darnton-Hill and Nalubola (2002) identified the "support of industry, with early involvement of local industry and the private sector" (235) as one of the key success factors for fortification initiatives. The partnership with the private sector is a dominant reason for the enthusiasm for fortification by the Business Alliance for Food Fortification (BAFF) with multinational corporations from Coca-Cola to Nestlé. Biofortification emerged in the context of the growing need of international agricultural research institutions to draw on corporate expertise and resources (Brooks 2010).

8. While the origin of the Green Revolution can be located with private foundations such as the Rockefeller Foundation, which provided funding to improve yields of corn, wheat, and beans in the 1940s, the agricultural research centers of private foundations were eventually consolidated under the Consultative Group on International Agricultural

Research (CGIAR), which was a loose network of national centers with various funding sources, and the network of national agricultural research systems in different countries. In addition, governments took the lead in promoting the Green Revolution (Gupta 1998). In Indonesia, the government conducted the Mass Guidance program (BIMAS) that distributed necessary agricultural inputs, particularly in Java (Hansen 1978). They were accompanied by a food price–control mechanism via the Food Logistics Agency (BULOG) (Arifin 1993; Thorbecke and van der Pluijm 1993).

9. Neoliberalization describes the rise of neoliberal ideology but pays attention to its heterogeneity and open-endedness, as opposed to the "teleological reading of neoliberalism" (Peck and Tickell 2002, 400).

10. See Avakian and Haber (2005) for a summary of works on women and gender, mainly in anthropology, history, and cultural studies. For pioneering gender work in rural sociology and geography, see Sachs (1983; 1996) and Whatmore (1990), among others.

11. According to the UN's Food and Agriculture Organization (FAO), on average, 43% of the agricultural labor force of developing countries is female. The female share of the agricultural labor force varies widely from 20% in Latin America to almost 50% in East Asia, Southeast Asia, and sub-Saharan Africa (FAO 2011, 7).

12. Note the peculiarity of the pattern of women's incorporation into the global food production system. An increasing number of women work as hired laborers, but women are less likely than men to own land and livestock (FAO 2011). For more on the feminization of agriculture, see Barndt (2002), Barrientos (1997), Carr, Chen, and Tate (2000), and Raynolds (1998).

13. More problems arise from the fact that many of the new types of export crops—such as vegetables—are traditionally considered "women's crops" in many parts of the world. New export crops often mean that women's plots are put under the control of men (Carney 1994; Dolan 2001). For more discussion on the gendered nature of contract farming, see Dolan (2001), Raynolds (2002), and Carney (1994).

14. The works in this area are too many to provide a comprehensive list, but classics include Amartya Sen's *Poverty and Famines: An Essay on Entitlement and Deprivation* (1981), Alex de Waal's *Famine That Kills: Darfur, Sudan* ([1989] 2005), Mary Howard and Ann Millard's *Hunger and Shame: Child Malnutrition and Poverty on Mount Kilimanjaro* (1997), and many World Watch Institute works on food issues such as *Underfed and Overfed: The Global Epidemic of Malnutrition* (Gardner and Halweil 2000).

15. For instance, the late-nineteenth-century food reform movement in the United States attracted middle-class women who later became active in local schools and charity organizations. "Domestic science" gave women from a privileged background the opportunity to gain higher education and a respectable career (Shapiro 2009). World War II's food programs elevated women's status by praising their patriotic contribution to the war effort (Bentley 1998). More recently, alternative agrofood movements that aim to create a more sustainable food system have attracted women activists (Allen 2004; Allen and Sachs 2007; DeLind and Ferguson 1999). Women are also overrepresented in contemporary food education movements, and some women are able to gain access to public policymaking as experts on food education and food literacy (Kimura 2011).

16. Laura Shapiro, in her historical analysis of the cooking school movement, provides various examples of irony in the food reform movement. For instance, she quotes a newspaper column that said of then-increasing labor protests and strikes: "Many of the so-called strikers would strike no matter how much work they had on hand" and "They are illy fed. Not from lack of money, but from lack of knowledge. Poor things, how are they to find out the best food to sustain their needs? ... I verily believe if the rigid instructions for food and feeding were implanted in the minds of our girls during their early school days, the labor element would not be such discontented individuals" (2009, 131).

17. For instance, in Indonesia, the New Order government of Suharto, which replaced the Sukarno government in the 1965 coup, wholeheartedly embraced the paradigm of "overpopulation." The government set up the Family Planning Institute (Lembaga Keluarga Berencana Nasional) in 1968, which in two years became the National Family Planning Coordinating Board (BKKBN). The BKKBN was very powerful, being operated directly under the president's supervision (Achmad 1999), and well funded, and it amassed a large workforce (Achmad 1999; Caldwell and Caldwell 1986). Contraceptive devices and pills were well stocked by the central government. The BKKBN also had an extensive network at the village level, employing many Family Planning Field Workers (Petugas Lapangan Keluarga Berencana) in local municipalities. These Family Planning Field Workers were the arms of the government, vigorously promoting contraceptive use, sometimes in a coercive manner (Achmad 1999; Hull and Hull 2005; Newland 2001).

18. During the Green Revolution, increase in agricultural yield through modern technologies was seen as a critical ingredient for national development by many Third World leaders. Cullather (2004) argues that "developmental populists couched the goal of self-sufficiency of food in nationalist terms, as an attribute of a progressive, independent nation" (246), pointing out that two slogans of Ferdinand Marcos in the Philippines were "Rice, Roads, and Schools" and "Progress Is a Grain of Rice," while one of Dudley Senanayake's in Ceylon was "Grow More Food." The Green Revolution's "miracle grains," he argues, became "a living symbol of abundance, an apparition capable of inducing mass conversions to modernity" (228).

19. Within agrofood studies, one can find excellent analyses of the political economy of the agrofood system, particularly its globalization, in works such as Bonanno et al. (1994), Friedmann (1991), and Magdoff, Foster, and Buttel (2000). For gender analysis of globalized food, see J. Collins (1993; 1995) and Barndt (2002). Many works have analyzed the cultural and social history of food through commodity case studies, including Pilcher (1998) on corn, Grossman (1998) on bananas, DuPuis (2002) on milk, and Dixon (2002) on chicken. Agrofood studies have also documented and encouraged recent social mobilizations that problematize the status quo of policies and structures of modern food. Some conceptual developments influential in food movements include Kloppenburg and Lezberg's (1996) call for localizing the "foodshed," Lang and Heasman's (2004) and Lang's (1999) call for "food democracy," and Lyson's (2004) call for "civic agriculture." Critical analyses of food movements include Allen (2004) and Hinrichs and Lyson (2008).

20. In an approach pioneered by Friedmann (1982; 1987), Friedmann and McMichael (1989) theorized two distinct food regimes. In the first food regime (1870–1914), the New World supplied cheap food to Europe, which lowered wage costs and supported extensive capital accumulation. The second food regime (1947–73), formed under US hegemony, constituted a livestock complex and a wheat complex. Bringing together insights from regulation theory and world systems theory, food regime analysis pays particular attention to international food complexes and how they are linked with key changes in the state systems. It points out symbiotic relationships between capitalism and particular configurations of food relations. For elaborations and extensions of these authors' work, see the 2009 special issue of the journal *Agriculture and Human Values* (Campbell and Dixon 2009).

21. Instead of direct income support, the US government opted for price supports for agricultural commodities, which meant that it needed to control imports and subsidize exports. Food aid was a key mechanism of subsidized exports that did not lower world market prices (Friedmann 1993, 33).

22. Since the early 1950s, world wheat exports have increased 2.5 times; the US share has increased substantially (Friedmann 1982).

23. Chapter 5 examines how this happened in Indonesia. Another example of the profound impact of US wheat is in Japan, where the United States encouraged consumption of wheat through numerous trade missions and school lunch programs. As a result, Japan "became the largest of the new wheat importing countries after World War II" (Friedmann 1982, 43).

24. For instance, Pritchard (2009, 299) points out how, during the first food regime, India exported grains but in the 1960s became dependent on US grain imports, absorbing up to 25% of the annual US wheat crop in some years in the 1970s. However, the Green Revolution reduced the necessary imports, and by the 1990s, India became a net exporter of grains.

25. For instance, Indonesia's BIMAS program for rice intensification used foreign companies such as the Swiss chemical company, Ciba, the German chemical company, Hoechst, and the Japanese trading company, Mitsubishi, to distribute agricultural inputs. Crouch (2007) argues that the BIMAS "led to a substantial increase in rice production through the introduction of new seed varieties, but it was also very profitable for companies involved which were guaranteed payment by the government" (290).

26. McMichael (2005) argues that this privatization of security under the globalization project is profoundly different from socialization of security under the development projects.

27. The WTO's Agreement on Agriculture was negotiated in the 1986–94 Uruguay Round. It aimed to improve market access and reduce trade-distorting subsidies in agriculture.

28. The WTO required all states to allow imports of at least 5% of domestic consumption (McMichael 2005, 277).

29. I put "cheap" in quotation marks because they are often artificially cheap. According to the Institute for Agriculture and Trade Policy, in 2003 US wheat was exported at an average price of 28% below the cost of production, corn at 10% below, and rice at 26% below (Hansen-Kuhn 2011).

30. Describing global agricultural trade, McMichael (2005) points out that the political function of the notion of privatized food insecurity was to add further justification for pressing markets in the global South to open up to products from the global North. The neoliberal mantra of the free market obfuscates the reality of the political determination of the market. The hypocrisy of the "free trade" regime is that powerful countries, notably the United States and the EU countries, continue to subsidize their agriculture, whose artificially cheap produce floods developing markets. The notion that trade can reduce food insecurity has helped to justify opening up states in the global South, while the states of the global North have managed to keep their subsidies thanks to their political and economic advantages (Rosset 2006).

31. For excellent discussions of biopower and modern science, see Dan-Cohen and Rabinow (2006), Kay (2000), and Petryna (2002). Scholars have also used the concept productively in relation to the developing countries; see Escobar (1995), Peluso and Vandergeest (2001), Agrawal (2005), Goldman (2001), Gupta (1998), Shivaramakrishnan (2003), and Anderson (2002).

32. Examples of such interventions abound in history. For instance, the bodily conduct of local subjects in matters of hygiene and nutrition has provided humanistic justification for continuing the civilizing mission of colonial power (Anderson 2002; Arnold 1993). Scholars analyzing contemporary international development projects similarly have found that the representation of the "problems" of Third World peasants, women, or the environment have helped justify additional development projects (Escobar 1995), enabling those from the developed North to portray themselves as educated, modern, enlightened, and

benevolent (Mohanty 1991). Identifying "problems" has never been innocent of political consequence.

33. A good example of the difficulty of defining critical human needs is the failed attempt by international development experts to define and standardize "basic human needs." Starting in the 1960s, international development organizations such as the International Labor Organization (ILO) and the FAO tried to define "basic human needs." Yet the endeavor, in spite of much excitement and investment, was ultimately not successful. It was reduced to either "specify what commodities fulfill basic needs and so run the risk of introducing culturally unsuitable goods" or "provide abstract definitions that are virtually unusable" (Douglas et al. 1998, 213). As Mary Clark put it, "the abstract word, 'needs', is never clearcut" (quoted at 206).

34. Feminist scholars have pointed to the politics of expertise in defining human needs. In her analysis of the welfare state, Nancy Fraser (1989) examines the construction of women's "needs" according to the specific logic of a managerial bureaucracy. With the notion of "politics of needs interpretation," she highlights the ways in which potentially political "needs" are depoliticized and subsequently naturalized. The politics of needs is also a subject of Haney's (2002) analysis of Hungary's welfare system.

35. The assumption that food governance is best left to "experts" also relates to a broader cultural understanding of a lay-expert divide that sees laypeople as incapable of understanding technical issues (see, e.g., Brooks and Johnson 1991; Perhac 1996). Many theoretical and empirical studies show how this might not be the case. The involvement of laypeople in a formerly expert-only space has grown in research on HIV/AIDS (Epstein 2000), breast cancer (Brown et al. 2006), and environmental pollution (Brown and Ferguson 1995). There have also been experiments to establish a forum to bring lay citizens into policy debates of a highly technical nature. Citizen involvement in the deliberation of technoscientific matters in the forms of "citizen jury," "science café," and "consensus conference" have been tried in many areas from biotechnology and telecommunications to nanotechnology (Powell and Kleinman 2008; Rowe and Frewer 2005; Sclove 2000).

36. The decline in public international agricultural research has been accompanied by the expansion of private sector research. Since the 1990s, for instance, private breeding programs have superseded public breeding programs and 38% of agricultural biotechnology patents are held by five private corporations (Byerlee and Dubin 2009). Nestle (2002) also documents how industry interests shape the direction and agenda of nutrition research.

37. Additionally, Brooks (2011) notes how the new cohort of private charity organizations such as the Bill and Melinda Gates Foundation emphasizes science-based solutions and "break-through science." This suggests the implications of scientized framing of food insecurity and the lure of technical fixes.

38. The earlier productivist policies of the Green Revolution also scientized food insecurity. Gupta (1998) observes that the Green Revolution was envisioned as "the application of 'scientific methods' and a top-down, production-based strategy" (53) and that it operated with the assumption "that scientific work inherently results in the greater social good" (56).

39. The Bill and Melinda Gates Foundation committed a total of $14.7 billion to global health and $1.8 billion to agricultural development between 1994 and 2011. The largest agriculture-related grants by the Gates Foundation were all started after 2006. It committed $100 million to the Alliance for a Green Revolution in Africa in 2006, $45 million to HarvestPlus II for biofortification in 2008, $33.3 million to the International Maize and Wheat Improvement Center's project on drought-tolerant maize for Africa in 2006, and $39.1 million to the African Agricultural Technology Foundation's project on water-efficient maize in 2008 (Bill and Melinda Gates Foundation 2011).

40. For instance, the WFP is still pushing fortified food, and CGIAR in 2008 identified biofortification as one of the "best bets" worthy of "scaling up" (Brooks 2011).

41. The demand for biofuel has increased since 2003 and consumed 25% of US crops in 2007. High oil prices are a factor, not only because they affect fertilizer prices but also because they make biofuel prices competitive. Poor harvests of US and Australian wheat were also a potential contributing factor. In addition, important rice-exporting countries such as Vietnam and India banned the export of rice in 2007–8. China's and India's increasing appetites, especially for meat, are also thought to be a factor, although the increase in demand has been steady. What's more, China and India are importing less wheat than in the 1990s, and India is generally a net exporter of rice (Headey and Fan 2008).

42. While various governments have also declared renewed commitment to agriculture, they tend to view the role of the public sector as laying the basis for the private sector. For instance, the African Union urged its member countries to increase public investment in agriculture by a minimum of 10% of national budgets to increase agricultural productivity by at least 6% in the Comprehensive Africa Agriculture Development Programme in 2003, and this was followed by the Abuja Declaration at the Africa Fertilizer Summit (Miltz 2011). These government activities are also accompanied by a belief in the power of the private sector, as evinced by a report from the Fertilizer Summit that says that "the underlying thesis was that an enabling environment must be created for the identification of actionable programs that, if implemented, will result in the establishment of private sector-led fertilizer markets to achieve the African Green Revolution" (Wanzala and Roy 2007, 2).

43. Miltz (2011) discusses how subsistence farmers are often forced to convert to monocropping and intensive use of agrochemicals and cites a farmer who said that "the authorities wanted us to become commercial seed growers, but the women of the cooperatives wanted to keep growing sweet potatoes, cabbage, and other vegetables in the marshes. They wouldn't back down and the authorities wound up sending in the army to pull up our crops."

44. A 2008 article in the journal *Foreign Affairs*, entitled "The Politics of Hunger: How Illusion and Greed Fan the Food Crisis," by an Oxford University economics professor is another example of the four shared characteristics of nutritionist and productivist discourses. In his polemical analysis of the crisis, Paul Collier condemns the "middle- and upper-class love affair with peasant agriculture" (71) and "romantic hostility to scientific and commercial agriculture" (73) for the making and exacerbation of the crisis. He argues for the increase in production of food as the key solution, saying that "the world needs more commercial agriculture" and "the world needs more science" (68). Specifically, he argues that in order "to counter the effects of Africa's rising population and deteriorating climate, African agriculture needs a biological revolution" (76). In his view, peasants "are ill suited to modern agricultural production" (71) and are incapable of innovation and entrepreneurship. Arguing that "peasants, like pandas, show little inclination to reproduce themselves" (70), he embodied the view that the problem lies in the lack of science and modern technologies and that the solutions would come from politicians, scientists, and the private sector, not from the urban/rural poor themselves.

45. A growing number of scholars have noticed that "quality" has become a central organizing principle in contemporary food systems. While governments are increasingly hesitant to impose standards, given their WTO commitments, private corporations, especially retailers, are imposing their own standards (Busch 2011; Friedmann 2005). Social movement–inspired standards such as organic, fair trade, and animal welfare labels have also proliferated. Several studies have pointed out that demands for higher quality often get appropriated as private standards, expanding corporate control and increasing profit margins (see, e.g., Guthman 2007; Kimura 2010; Mutersbaugh 2005). I point out a

different yet related politics of food quality and how it intersects with changing notions of food insecurity. Friedmann noted that, with the growing power of private standards regulating global sourcing of food, two types of food are provided by corporations: rich consumers get "fresh, relatively unprocessed and low chemical input products" assured by privatized quality assurance systems whereas poor consumers get cheap, standard commodities (2005, 258). On the surface, the quality turn might seem to contradict such an analysis, given that the majority of people in the developing world fall into the latter category. However, by framing food insecurity in strictly micronutritional terms, cheap standard commodities can be framed as quality products as long as they have added micronutrients. The attention to "quality" often has functioned to justify the use and benefits of biotechnology and "durable food" in the developing world.

46. Plumpy'nut is a French patented fortified peanut paste that has been used by UNICEF and other aid organizations to address acute malnutrition. The *New York Times* (Rice 2010) reported that its American manufacturer wanted to expand the market base by using it to prevent malnutrition. GSK sells health drink and fortified instant noodle under the Horlicks brand.

2. CHARISMATIC NUTRIENTS

1. Of course, vitamin A is a type of micronutrient. I consider vitamin A here as a separate charismatic nutrient, since it was not until the 1990s that the overarching concept of "micronutrients" became a focus of concern.

2. Medical anthropologists have used the term "charismatic authority" to describe non-Western healing in contrast to the "rational" authority of Western medicine (see, e.g., MacCormack 1981; 1986). My intention here is to flip this argument over and direct a similar gaze at Western medicine.

3. McLester continued to link superior physique with protein intake in the 1949 edition of the book, arguing that "the development of races as well as that of individuals may be influenced by the liberality of the intake of protein" (60).

4. For instance, Waterlow later reported on the same disease but refused to call it kwashiorkor. Instead, he called it "fatty liver disease" (Ruxin 1996, 67).

5. They identified fish, soybeans, peanuts, sesame, cottonseed, and coconut as the ideal candidates.

6. Pretorius and Smith (1968) found that when children with kwashiorkor were fed diets with relatively low protein but high energy content, they showed healthy recovery. In 1975 Philip Payne found that any diet with a density of protein greater than about 10% would be sufficient for nutrition and that most staple grains meet this criterion (cited in Solomons 1999, 154).

7. Partly because of frustration with protein's ineffectiveness, international organizations started to look for alternative health programs. For instance, UNICEF decided to focus on growth monitoring, oral rehydration therapy, breast-feeding, and immunization to improve child survival. This was decided at the 1977 Alma Ata meeting and called the GOBI initiative. Lindsay Allen (2003) argues that the central impetus for these activities was the growing concern that little progress was being made in addressing protein deficiency.

8. Their difficult balancing act is also reflected in the following statement by the FAO/WHO, which was at pains to assert the legitimacy of the existing protein paradigm while recognizing the need for modification: "As the widespread use of the term marasmic kwashiorkor suggests, kwashiorkor may be superimposed on any degree of marasmus . . . the attention of these investigators and of those responsible for preventive and corrective programmes should be directed, without decreasing the interest in kwashiorkor, to all aspects of the problem of protein-calorie-deficiency disease" (quoted in Carpenter 1994, 182).

9. Indeed, one might say that the story of protein is a story of women made invisible. It was a woman whose kwashiorkor research was long marginalized. Later, many protein projects implicitly took women as only a means to an end.

10. According to Reddy (2002), USAID established the International Vitamin A Board in 1973, and USAID and WHO held a joint meeting in Jakarta that led to the establishment of IVACG in 1975. IVACG included not only academic researchers but also policy-makers and development practitioners. IVACG holds regular meetings to discuss vitamin A issues that emerge both in academia and the practical policy arena.

11. The interview with Sommer took place in Baltimore on September 7, 2004. A meta-analysis of replicate studies confirmed the Aceh study's claim, finding on average a 23% reduction in mortality in children (Beaton et al. 1992).

12. By the 1960s, vitamin A's role in the prevention of xerophthalmia was known thanks to several studies (see, e.g., Oomen, McLaren, and Escapini 1964). WHO conducted the first global survey of xerophthalmia in the early 1960s.

13. Interview with Alfred Sommer, September 7, 2004.

14. However, it should be noted that the breast milk of the poor Indian mothers they studied had lower levels of vitamins.

15. It is instructive to note what the decoupling of this dyad does to women. In addition to the neglect of "non-reproducing" postmenopausal women's issues, another stark example is provided by Kilaru et al. (2004) in their analysis of Indian policy on reproductive health. They point out that the attention to children's health led the government to focus exclusively on medical risks during pregnancy and at childbirth, while postpartum risks were not sufficiently acknowledged. And this, despite maternal death being more common during the postpartum period than in the prenatal period or childbirth itself. Once separated from the child women's medical and policy value diminishes.

16. Historian Ann Stoler (2002) points out the particularly complicated history of native women's motherhood that was tangled up in colonial anxiety about racial purity and sexuality.

17. Home economists in the United States in the early twentieth century were predominantly (white, middle-class) women. Many of them entered the field because they could not enter other natural science fields (Levine 2008, 19).

18. The concept of a "boundary object" highlights the importance in scientific work of materials or objects that facilitate linking of researchers by serving as a focus of attention and forming a basis of shared identity (Star and Griesemer 1999). Joan Fujimura (1992) has proposed another useful concept, that of a "standardized package" of technologies, which enables researchers to construct a relatively stable object of research and standardized methods of investigation.

3. SOLVING HIDDEN HUNGER WITH FORTIFIED FOOD

1. Biofortification is a new addition to this list of "solutions."

2. Of course, economization of nutrition and health is not limited to the practice of international development. For an example of economization of health policies in developed countries, see Sjögren and Helgesson 2007.

3. The 1991 "Ending Hidden Hunger" conference specifically addressed micronutrient deficiencies and ways to combat them. At the 1992 International Conference on Nutrition, representatives from 159 countries reaffirmed the goals of the 1990 World Summit for Children.

4. The Program Against Micronutrient Malnutrition is coordinated by the faculty at the Rollins School of Public Health of Emory University, the Centers for Disease Control and Prevention (CDC), and program officers at the Task Force for Child Survival and Development. PAMM's network also extends to the International Agricultural Center and

the Department of Human Nutrition of Wageningen Agricultural University in the Netherlands (OMNI n.d.).

5. PubMed is a database of the National Library of Medicine, which includes over 14 million citations for biomedical articles back to the 1950s. The URL is http://www.ncbi.nlm.nih.gov/entrez/query.fcgi.

6. Initially, corn-soy milk from the World Food Programme and USAID was distributed to children. But international donors decided to look for fortified products and worked with the government to identify locally made baby food products. They chose an Indonesia-based food conglomerate, Indofood. Indofood is one of the largest food manufacturers in Asia and produces a variety of food products, most notably, instant noodles. Indofood already had a popular product, SUN baby food, that was fortified with micronutrients. In April 1998, the phase 1 distribution of SUN started as a national emergency nutrition intervention. To promote its acceptance, the government and donors conducted a social awareness campaign with the slogan "Save young children from being a lost generation." "The lost generation" became a popular phrase to talk about the impact of "hidden hunger" in the country. The SUN emergency distribution originally targeted select areas considered most affected by the crisis: slum areas near four cities where many factories were closed down and many women workers were suddenly unemployed. Subsequently, UNICEF campaigned among its donors to expand the project to other areas, and major donors such as Australia, Norway, the UK, and Canada agreed to fund it.

7. Vitadele was also produced by Indofood.

8. Interview with WFP staff in Jakarta, October 2004.

9. Interview with Mercy Corp, Indonesia country director, December 2004.

10. Through International Relief and Development, the USDA donated rice, soy, and soy flour, which local contracted factories made into these various products.

11. Interview with staff at Helen Keller International, Indonesia, November 2004.

12. It has fourteen vitamins and minerals, providing one-eighth of the recommended daily allowance (RDA) for children under five years old.

13. See the Grameen Group website, www.grameen-info.org.

14. It is interesting that a fortification program in a given country often starts without a complete set of data so that it can ascertain whether the program improves the nutritional status of a target population as intended. A powerful discourse that has rationalized fortification involves the notion of "mimicry" of the West. Akhil Gupta has argued that the broader notion of "Third World development" maps developing nations as juniors in relation to the West, with the key to overcoming such junior status being to mimic their senior by learning to "follow, replicate, repeat, improve" (1998, 40). The call to follow the West's example has figured powerfully in the global debates on fortification, and its proponents have drawn on fortification experiences of developed countries. For instance, the World Bank's *Enriching Lives* (1994) promoted fortification by basing its recommendation on the *assumed efficacy* of fortification in the developed nations. This kind of reasoning was most clear in a section titled "How Fortification Won in the West" that emphasized the effectiveness of fortified flour in decreasing anemia in the United States and vitamin D-fortified margarine in eradicating rickets in Britain (World Bank 1994). It further proclaimed that "indeed, fortification ... has eradicated most vitamin and mineral deficiencies in the industrial countries" (27). Such a line of logic is not limited to the World Bank. UNICEF also promoted fortification by claiming that "fortification of an appropriate vehicle with specific nutrients has been practiced in numerous industrialized countries for many years with considerable success" (Darnton-Hill et al. 1999, 26). Evident in these statements is the assumption that developing nations are temporarily inferior to the West and that "however the paths or strategies to achieve development are described, the means to that end is assumed to be mimicry" (Gupta 1998, 40).

15. Actual impact of SAPs is highly contended. For instance, a 1998 study by the World Bank on health expenditure found little negative impacts from the SAP projects (Ruger 2005).

16. The 1984 figure is from Fair (2008). In 2007, the health and social service sector ($2.8 billion) constituted about 11% of total World Bank lending ($24.7 billion). Other large sectors included law, justice, and public administration (22%); transportation (20%); and water, sanitation, and flood protection (12%) (World Bank 2008a).

17. The economization of nutrition was not limited to the World Bank in the 1990s. Nutritional experts have tried various arguments to try to get attention from funding agencies and governments. Casting malnutrition in terms of the economic losses it causes is one of the arguments that some international organizations have tinkered with. For instance, in the 1960s, the FAO made a similar argument when it tried to recast malnutrition as the cause of worker lethargy and loss of productivity (1962). The Protein Advisory Group also experimented with such economic language when in 1965 it said, "The maimed survivors become adults lacking the vigor and enterprise essential for productive advancement. Their shortened life span and decreased ability to produce gravely impede the physical, mental and economic growth of the population" (quoted in Ruxin 1996, 179). Yet, since the 1990s, the economization of nutrition has been promoted by powerful actors who exert tremendous financial and epistemological influence in the sphere of international development.

18. It should be noted that in 2008 major grain-trading companies saw record profits at a time when the poor suffered from high food prices. The *Wall Street Journal*, in an article titled "Grain Companies' Profits Soar as Global Food Crisis Mounts," reported that Archer-Daniels-Midland's profits jumped 42% and Cargill's 86% (Kesmodel, Etter, and Patrick 2008).

19. It is telling that World Bank president Robert Zoellick continued to promote the market as the solution for the food crisis even during the 2008 financial crisis. He stated that global trade was the "key to lower food prices" and thus was contributing to, rather than destabilizing, food security (World Bank 2008b).

20. This study, published in the *American Journal of Clinical Nutrition*, followed thirty-three Indonesian women who were instructed to take iron tablets every day for two months. They matched the results of tests on the women's stool samples and the women's claims to have taken all the pills and found that "although 64% of the women claimed to have taken all iron tablets, the actual percentage of women who took all tablets is most probably much lower" (Schultink et al. 1996, 137).

21. The concept of gender mainstreaming was officially adopted at the Fourth World Conference on Women in Beijing in 1995, which committed UN organizations to systematically incorporate a gender perspective into policymaking. It is now the official policy of the UN, and many governments have adopted the concept. For discussion of the role of transnational networks in promulgating the concept worldwide, see True and Mintrom 2001. Gender mainstreaming should be located in a longer history of struggles to get the international development community to pay attention to women. The 1970s concept of "women in development," or WID, was a response to criticism for neglecting the role of women in development. Although initially welcomed as an improvement over the previous neglect of women's role in development, scholars have criticized WID for understanding "women" as a generic category devoid of history and culture. WID was also criticized for offering a restrictive understanding of the transformative capacity of women, as women were considered victims of men. An alternative framework, "gender and development," has been proposed to discuss the social construction of gender and the interconnections of gender, class, and race (Parpart, Connelly, and Barriteau 2000).

22. One of the problematic implications of Brown's contention is that the rejection of such identities might make social activism less effective. Whether "identity politics" is

necessary or desirable has been contested fiercely by feminist scholars (Scott 1996; Fraser 1997; Young 2000; Butler 2000; Pratt 2004), as there is a recognition that identity-based claims ("we women") could generate "subversive energy" for feminist organizing (Pratt 2004, 71) while still risking universalizing a subcategory of the group (such as white, middle class, or heterosexual). The problem of biological victimhood goes further, in that its claiming is not made by women but imposed by scientific and technocratic experts.

23. In her analysis of international health politics, Beall (1997) notes similar dynamics when women, particularly mothers, are identified as the strategic point for health interventions that aim to improve key health statistics such as maternal mortality rates and infant mortality rates. But, she writes, "the result has been the use of women as development solutions, to increase the effectiveness of development interventions, rather than to accord them any agency" (82).

4. BOUND BY THE GLOBAL AND NATIONAL: INDONESIA'S CHANGING FOOD POLICIES

1. PAMM is a project of the Rollins School of Public Health at Emory University, the US Centers for Disease Control and Prevention, and the Task Force for Child Survival and Development.

2. Micronutrient programs had existed before the 1990s. Indonesia's first major policy regarding a micronutrient was vitamin A capsule distribution, started with Helen Keller International's distribution of vitamin A capsules to children under five in Java (Pollard and Favin 1997). The government made it a national program in 1974. In the same year, the government started to distribute iron tablets to pregnant women in order to combat iron deficiency anemia (Hartini et al. 2003). In addition, in order to tackle iodine deficiency disorder, the government started lipiodol injections of school children and newly married women in highly endemic areas. However, the implementation of these micronutrient-related programs was at best uneven. For example, even the vitamin A program, which was considered the most successful among these, reached less than half of the target population in 1986 (Pollard and Favin 1997). Moreover, these nutrition programs were dwarfed by expenditures on population and agriculture programs that tried to decrease the rate of population growth and increase agricultural production. The all-powerful population control machinery gradually seeped into what was technically a nutrition program. For instance, the Family Nutrition Improvement Program (Usaha Perbaikan Gizi Keluarga or UPGK) eventually came under the control of the powerful National Family Planning Coordinating Board, with many projects shifting from nutrition to contraceptive acceptance (Pandi 1987; Achmad 1999; Rohde 1993). So, while micronutrients programs did exist in the pre-1990 period, it was only in the 1990s that they started to command a growing presence in how food insecurity was framed in Indonesia.

3. Repelita is an acronym for Rencana Pembangunan Lima Tahun (Five-Year Development Plan) and a pun on *pelita* (lamp, or light).

4. This phenomenon is not limited to Indonesia. Kiess et al. (2000) summarize the general situation: "Traditionally, surveillance systems have relied on anthropometric indices of children to monitor health and nutrition ... there is little experience in incorporating indicators of micronutrient status, such as anemia and vitamin A deficiency, into surveillance systems and interpreting the trends and patterns of such indicators" (230).

5. Interview with a former BAPPENAS staff member, November 2004.

6. Interview with a staff member at the Ministry of Health, April 2005.

7. The KFI was established with Soekirman (formerly with the government's BAPPENAS), Suroso Natakusuma (formerly with government agencies BULOG and the Office of State Minister of Food Affairs), and Thomas Darmawan from the food industry as founders. It also includes industry members such as the CEO of Bogasari Flour Mill and the CEO of Kimia Farma.

8. Some Indonesians have only one name and others have a first and last name.

9. Repelita VI (1994-99) states: "Greater attention will be paid to efforts to overcome the problem of IDD, remembering its negative impact on children's intellect and psychology. For that purpose, the addition of iodine to salt [salt iodization] for consumption will be conducted" (Repelita VI, 188) (Government of Indonesia 1993).

10. In addition, as we saw in chapter 3, fortification was portrayed as a perfect example of private-public partnership, the use of the market approach, and a cost-benefit–efficient public policy. The noted nutritional expert Soekirman echoed this widely circulated discourse by insisting that the "food industry community has to be the pioneer and the primary actor" for fortification, and he urged the industry to become aware that fortification would provide additional income (Soekirman 1998, 913). With fortification, the solution to the micronutrient problem could be offered through the market rather than the state, and this was good for national development.

11. Puslitbang Gizi was first established by the colonial Dutch government as the Institute of Nutrition Research in 1934. It conducted nutrition research, surveys, and education, as well as advising the government (Soekirman et al. 2003). After independence, the institute was renamed Lembaga Makanan Rakyat (LMR) under the leadership of Poorwo Soedarmo, the father of nutritional science in Indonesia. In 1967, it was split into two bodies—one became the policymaking body (the current Directorate of Community Nutrition in the Ministry of Health), and the other became the research body, initially directly under the Directorate of Community Nutrition as Balai Penelitian Gizi and in 1975 renamed Puslitbang Gizi and with a higher bureaucratic status. Many prominent nutritional scientists have been affiliated with Puslitbang Gizi. For instance, Muhilal and Darwin Karyadi, who both have served as its head (Karyadi: 1975–93; Muhilal: 1994–99), are prominent nutritional scientists whose works have appeared in leading Western academic nutrition journals.

12. Interview with a staff member at the Food and Nutrition Research and Development Center, August 2005.

13. Interview with a staff member at Bogor Agricultural University, July 2005.

14. decree is No. 202/BM/DJ/BGM/II/1996 Tanggal 13 Februari 1996 tentang Penanggulangan Anemia Gizi bagi Pekerja Wanita (Direktorat Bina Gizi Masyarakat 1996).

15. Spar (1996) reports that in1995 shoe and textile industries opposed the increase in the Indonesian minimum wage, saying that it would force them to move their business outside the country (35).

16. The Marsinah case galvanized labor activism. The mid-1990s saw an increase in labor protests, often led by women workers (Silvey 2003).

5. BUILDING A HEALTHY INDONESIA WITH FLOUR, MSG, AND INSTANT NOODLES

1. It reads "Certificate of Appreciation, Presented to PT Sriboga Raturaya, for being the First company in the world to Fortify its wheat flour with Zinc in addition to iron, thiamin, riboflavin and folic acid. The United Nations Children's Fund (UNICEF) expresses its appreciation to the company for its concern for improved health and nutrition as part of this business mission and for taking on this noble initiative on a voluntary basis" (Woodhouse 1999).

2. Interview with a former Office of State Minister of Food Affairs official, December 2004.

3. There was a corporate partner called the Zurich Group that helped open the flour mill, but I could not obtain further information on this group.

4. Bogasari also became a subsidiary of Indofood in 1995.

5. According to APTINDO (2003), the final use of wheat flour in the country is as follows: wet noodle and small industries (32%), instant noodle (20%), bakery and cake

(20%), household (10%), biscuit and snacks (10%), and dry noodle (8%). Given the increase in wheat consumption in Indonesia, researchers have discussed "Westernization of diet" (Fabiosa 2006). Many now consume wheat-based products, particularly instant noodles.

6. According to the USDA PSD data, in 2010, Indonesia was the world's third-largest wheat importing country, after Egypt and Brazil (USDA n.d.).

7. This effort was successful in 2003, raising the tariff back again from 0% to 5% by decree: Surat Keputusan Menteri Keuangan no. 127/KMK.01/2003 tanggal 10 April 2003 tentang Perubahan Tarif Bea Masuk atas Impor Tepung Gandum (Ministry of Trade and Industry 2003).

8. Interview with a former staff member at the Office of State Minister of Food Affairs, December 2004.

9. Interview with a staff member at the Ministry of Health, April 2005.

10. Kesehatan no. 962/Menkes/SK/VII/2003 tentang Fortifikasi Tepung Terigu. Before this regulation, there was another decree (SK no. 632/Menkes/SK/VI/1998 tentang Fortifikasi Tepung Terigu), but it did not require registration.

11. Of course, without mandates, the private sector has fortified the products with various vitamins and minerals since the 1950s (interview with a former employee of Indofood, December 2004).

12. Interview with a staff member at the Ministry of Health, April 2005.

13. Interview with a staff member at the Ministry of Health, April 2005.

14. Interview with a staff member at the Ministry of Health, April 2005.

15. Keputusan Menteri Negara Urusan Pangan no. Kep 14/M/08/1997 tentang pembentukan komisi fortifikasi pangan.

16. An interviewee in the Ministry of Health described the impact of this survey: "From the national survey, we found that we are facing not only macroprotein energy malnutrition, but also we find out that anemia is prevalent not only among pregnant mothers, but also for those under fives, as well as women of reproductive age. We see also deficiency of vitamin A not only among those under five, but also pregnant mothers and nursing mothers. In addition, I attended several international meetings, I see the global trend, that micronutrient intervention is very cheap, but very effective, if we implement in a proper manner. So it should be the national program" (April 2005).

17. There are some experts who are still advocating cooking oil fortification with vitamin A. They point out that even compared to sugar and wheat flour, more cooking oil is used by both rural and urban people. They further point out that the production of cooking oil and margarine are relatively concentrated (215 factories owned by 7 corporate groups), making it ideal for quality control. They also cite successes in other countries.

18. Interview with a member of the Indonesian Fortification Coalition, November 2004.

19. No. 632/MENKES/SK/VI/1998.

20. Interview with Muhilal, August 1, 2005.

21. The decree on SNI was No. 153/MPP/Kep 5/2001 no. 323/MPP/Kep/2001. The government issued several other decrees that facilitated the implementation such as Kepmen Menperdagan no. 153, Kepmen Menperdagan, no. 323/MPP/Kep/11/2001, and Kepmen Menperdagan no. 59/Mpp/Kep/1/2002 Jan 31.

22. USAID purchased the premix worth $850,000 from global chemical companies such as Roche and BASF. This initial premix only contained elementary iron.

23. Another science and technology studies scholar, Brian Wynne (2001), has similarly called for "critical self-reflexivity about the implicit limitations and contingencies of their own knowledge" in his criticism of scientific experts who tend to engage in "systematic patronization of the public as intellectually vacuous" (447).

6. SMART BABY FOOD: PARTICIPATING IN THE MARKET FROM THE CRADLE

1. To be specific, she fed the porridge to children less than six months old, which health authorities would consider a bad practice given the six-month exclusive breast-feeding rule.

2. Apple (1995) defines "scientific motherhood" as the idea that the practice of mothering ought to be based on scientific knowledge and guidance.

3. For instance, historians have found that scientific feeding and child rearing are linked tightly with the concept of modernity in the context of the West. See, for instance, Levenstein's discussion of the development of formula milk and how it was touted by scientists as "modern, scientific, and American" (2003, 128).

4. I obtained copies of all past issues from 1979 until 2005 at the publisher's Jakarta office. The initial plan was to sample three issues each year before 1989, but it turned out that older issues had very few advertisements for baby food (although many advertisements for books, medicines, and toiletries). Therefore, I analyzed all advertisements for baby food that appeared during the period 1979–89 to have a large enough sample. For a more recent analysis, I randomly selected three issues from 2005. All advertisements in these two groups (old and new) were translated and analyzed for significant concepts and themes. To be sure, there are some limitations to this methodology. *Ayahbunda*'s readership is very limited. Its price (currently Rp 17,500, approximately $2 per issue) makes it unaffordable for most Indonesians. Interviews with mothers indicated that they saw advertisements on television rather than in magazines. Therefore, the analysis of television commercials would have given a more realistic assessment of the exact messages that are received by the majority of consumers. However, as sampling TV commercials, let alone historical samples, was logistically difficult, I decided to take samples of advertisements in this magazine throughout its history. There is no strong reason to believe that producers drastically change messages from magazines to television. I also confirmed with mothers during the interviews that what they saw in TV commercials was similar to what appeared in the magazine in terms of main messages. In addition, because I am comparing the advertisements in the same medium (magazine) across time, I could expect that any possible biases, if they exist, would work in similar ways over time. The objective of the content analysis is to illuminate historical changes in the corporate marketing strategy and whether it exists in Indonesia's baby food industry, rather than to analyze class differentiation of marketing messages.

5. I use this word, "uncover," with an awareness of much debate among feminist researchers. (See, e.g., Sprague 2005 and Escobar 1995 for critiques of a hegemonic undertaking to make women "visible" in international development.) I hoped that interviews would provide a small window through which we could begin to imagine alternatives to nutritionism.

6. All the interviews were conducted in Bahasa Indonesia. Interviewees were contacted in the following way: I asked the Jakarta city municipality's health department to select four impoverished subdistricts. In each subdistrict, I relied on *kadres* (health volunteers) to select interviewees. Most of the women were stay-at-home mothers, although some of them occasionally sold homemade food on the street or from home. Informed by feminist research methods that key our attention to possible power relations between researcher and "subject" (Reinharz 1992), I was keenly aware of the social, economic, and cultural distance between myself (a Japanese academic) and my interviewees. Although I can never claim to have eliminated the power asymmetry between us, I tried to minimize it in various ways. First, I did not invite government officials to my interviews, although some officials strongly suggested that I should. Second, being aware of possible pressure from *kadres,* who tend to be older and somewhat better-off even though they were from the same neighborhood, I was able to ask the *kadres* to leave us alone. At the beginning of each interview, I also emphasized to the interviewee that the result of the interview

would not be reported to any authority, or to the *kadres,* and that I would not disclose her identity. With permission, I tape-recorded all interviews and later transcribed them. Most of the interviews were about one-hour long, and almost all of them took place at the interviewees' residences, or on streets in front of their houses. When other people were around, they sometimes joined the conversation.

7. It should be noted that once formula or solid food is introduced, it tends to decrease breast-milk production.

8. These are government-run community health posts and health centers that provide basic health services. The association with vitamins is likely to come from the fact that they serve as distribution points of vitamin A capsules.

9. Note the similar social function of fortified food products in the context of developed nations. One motivation for the food companies to market fortified food (or what is often called "functional food" in the advanced capitalist markets) is to defend themselves against growing criticism of the food industry for its neglect of the health impacts of its products, particularly for the rising rate of obesity (Heasman and Mellentin 2001). The industry strategy has been primarily to argue that "there is no bad food, there is only bad diet," pointing the finger at consumers' dietary choices rather than at their own unhealthful products (Oliver 2006). But with functional food, they can now proudly say that their products are "healthy." Such a strategy was clear in a new product from Coca-Cola (Diet Coke Plus), which was fortified and marketed as a "good source of vitamins." After a warning from the Food and Drug Administration, the company dropped the claim (Heavy 2008).

10. Peraturan menteri Kesehatan no. 240/menkes/Per/V/85 tentang pengganti air susu ibu dibidang peningkatan penggunaan air susu ibu di rumah sakit; revised in 1997 with Keputusan menteri kesehatan no. 237/Menkes/SK/IV/1997 tentang pemasaran pengganti air susu ibu.

11. Peraturan Pemenrintah Republik Indonesia no. 69/1999 tentang iklan dan lable.

12. For instance, interviewees talked about how clinic workers were given financial incentives by corporations to sell formula and baby food products.

13. In particular, see his discussion in the epilogue.

14. Hays (1996) points out that one of the central components of the ideology of "intensive mothering" is the need for expert-guidance. O'Reilly (2004) argues that "sacrificial motherhood" is characterized by the following themes: the need for care by the biological mother; the availability of the mother 24/7; the commitment of energy, financial resources, and time to the child; the prioritization of the child's needs before the mother's; and the need for expert instruction.

15. On the cultural construction of "good" versus "bad" mothers, see Chase and Rogers (2001), chap. 2.

16. It is also important to realize that some are considered "worthy mothers of the nation" while others are not. Patricia Hill Collins (1999) notes in the context of US policy that nation-states control mothers of different classes, races, and citizenship groups differently. It is the most marginalized who become the most visible and invite the most stringent forms of state control.

17. A similar case of class and racial stratification of the ability to comply with scientific guidance is documented by Litt (2000), who studied the impacts of medicalized motherhood in American minority communities.

18. Important books on the feminist debates on Foucault include Hartsock (1990), McNay (1992), Ramazanoglu (1993), Sawicki (1991), and Pratt (2004, chap. 2).

19. Lock and Kaufert (1998) similarly point out that women's response to medicalization is not a simple one of victimization but is characterized by pragmatism and ambivalence, as indicated by the title of their book, *Pragmatic Women and Body Politics.*

7. CREATING NEEDS FOR GOLDEN RICE

1. The top five solutions they chose included micronutrient supplements, combating malnutrition, the Doha development agenda, micronutrient fortification, and expanded immunization coverage. It is noteworthy that fortification was also chosen.

2. Beta-carotene is a precursor to vitamin A.

3. For instance, a search for the term "biofortification" in PubMed found the earliest entries were all after 2000 (Poletti, Gruissem, and Sautter 2004; Hossain et al. 2004; Timmer 2004; Bouis 2003; Bouis, Graham, and Welch 1999; King 2002).

4. Of course, criticism of the Green Revolution has much broader theoretical grounds. For instance, there has been significant epistemological criticism from ecofeminists such as Vandana Shiva (1988) and Shiva and Maria Mies (1993) that condemns the Green Revolution as an example of Western patriarchal violence.

5. There were some even earlier attempts. For example, a 1968 report by the Advisory Committee on the Application of Science and Technology to Development at the UN included the development of genetically improved plants as a potential tool to combat protein malnutrition (Carpenter 1994, 162).

6. For the Golden Rice research, the Swiss Federal Institute of Technology (1993–96) and the European Community Biotech Program (1996–99) provided funding along with the Rockefeller Foundation. Total research investment was $2.4 million over nine years (Potrykus 2004).

7. This is the title of a book by the president of the Rockefeller Foundation, Gordon Conway (1998).

8. Alston, Dehmer, and Pardey (2006, 322) write, "By 1970, the four founding centers—IRRI, CIMMYT, IITA, and CIAT—were allocated a total of $14.8 million annually. The progressive expansion of the number of centers, and the funding per center, during the next decade involved a 10 fold increase in nominal spending, to $141 million in 1980. During the 1980s, spending continued to grow, more than doubling in nominal terms to reach $305 million in 1990. The rate of growth had slowed but was still impressive. In the 1990s, however, although the number of centers grew—from 13 to 18 before contracting to the current 15—funding did not grow enough to maintain the level of spending per center, let alone sustain the growth rates."

9. The Gates Foundation became the wealthiest charity organization when Warren Buffett gave it $37 billion, increasing its endowment from $29 billion to $60 billion. According to Okie (2006), the foundation has committed more than 60% of its resources to health-related projects. In the mid-2000s, it started to increase funding for agriculture. For instance, the Alliance for a Green Revolution in Africa (AGRA) was started in 2006 (Toenniessen, Adesina, and DeVries 2008).

10. Total funding for HarvestPlus was $100 million. While this may sound modest in scope, it is about a third of the annual funding of the entire CGIAR system.

11. For instance, the World Bank pledged to rectify its neglect of the agricultural sector in its World Development Report 2008, and the Gates Foundation has also started to emphasize agriculture.

12. Currently, there are five "significant" developing countries in terms of GM adoption: Argentina, China, South Africa, India, and Brazil. These five countries account for 89% of GM crop area in the global South (ISAAA 2010).

13. For instance, with Argentine soybeans, 3% of the producers are responsible for 70% of the production (Binimelis, Pengue, and Monterroso 2009).

14. The vast majority of India and China's GM production is in Bt cotton. ISAAA (2012) reports that India had total of 10.6 million hectares in GMO production, of which 10.6 million hectares were in Bt cotton, while China had a total of 3.9 million hectares in GMO production, of which 3.9 million hectares were in Bt cotton, although there

seemed to be some plantings of GM papayas, tomatoes, and peppers. Brazil and Argentina are second and third in global soybean exports. (The United States is number one.) (Thoenes 2004).

15. The Network involves the International Rice Research Institute, the Philippines National Rice Research Institute (PhilRice), Vietnam Cuu Long Delta Rice Research Institute, India Department of Biotechnology, India Directorate of Rice Research, Indian Agricultural Research Institute, University of Delhi, Tamil Nadu Agricultural University, Patnagar University of Agricultural Sciences, Bangalore Chinsurah Rice Research Station, Bangladesh Rice Research Institute, China's Huazhong Agricultural University, Chinese Academy of Science, Yunnan Academy of Agricultural Sciences, and Indonesia Agency for Agricultural Research and Development (Golden Rice Humanitarian Board 2005).

16. Therefore, IRRI researchers have to find suitable varieties that are popular and successful in a particular environment to receive the new genes, such as BR 29 from Bangladesh, Immyeobaw in Burma, and Nang Hong Cho Dao and Mot Bui from Vietnam (Rice Today 2003). Golden Rice made from these *indica* rices has given disappointing results, however, with the carotene content at 1.05 mcg/g in the best line, lower than the original Golden Rice (1.6 mcg/g) (Datta et al. 2003).

17. In this survey, a variety of stakeholders—farmer leaders, business people, extension workers, and researchers—were asked whether they thought a particular biotechnology application is useful/risky/morally acceptable/to be encouraged. When asked what they thought about "use of modern biotechnology in the production of foods to make them more nutritious, taste better and keep longer," consumers, businessmen, extension workers, and farm leaders tended to think it should be encouraged. The nutritional application of GMOs received more positive responses compared to other applications, such as "taking genes from plant species and transferring them into crop plants, to make them more resistant to pests and diseases," "introducing human genes into bacteria to produce medicine or vaccines, for example, to produce insulin for diabetes," "modifying genes of laboratory animals such as a mouse to study human diseases like cancer," and "using genetic testing to detect and treat diseases we might have inherited from our parents."

18. Examples of similarly politicized foods include meat in Chile (Orlove 1997) and potatoes and chicken in Jewish culture (Frank 1985).

19. Belasco also points out that women have been another type of "victim" to be saved by technological fixes (2006).

CONCLUSION

1. Furthermore, the public health promotion of fortified processed food can accelerate the problem if ordinary consumers have difficulty distinguishing "properly fortified" from regular food. In the context of a growing global social movement against junk food in developing countries, the claim of "healthy fortified processed food" could be seen as part of a public relations campaign by the global food industry. For an example of the emerging anti–junk food movement in developing countries, see Consumer International (2008).

2. Food sovereignty's conceptual orientation becomes clearer when we compare it with its predecessor, "food security," as the food sovereignty concept was created as an explicit critique of the food security concept. "Food security" was defined at the World Food Summit (FAO 1996): "It exists when all people, at all times, have physical and economic access to safe and nutritious food which meets their dietary needs and food preferences for an active and healthy life." Although they are ostensibly both responding to the same problematic of the Green Revolution, there are several major differences between these two concepts. First, while food security is concerned with macro, aggregated food availability typically calculated on the national level, food sovereignty focuses on individual access to

food, particularly by marginal groups. Second, given its focus on national food availability, food security tends to see international trade in food as useful and helpful, whereas food sovereignty criticizes neoliberal trade policies. Third, food security demands that policy focus on food access in general and the purchasing of food. Instead, food sovereignty concentrates on access to and control of productive resources (Windfuhr and Jonsén 2005). In the case of Indonesia, food sovereignty activists acknowledge the existence of malnutrition and undernutrition in the country, but identify problems of the food system beyond a narrowed focus on the nutrient makeup of food. They point out that "farmers are often defeated by the concept of food security which is only emphasizing food availability even though it has to be imported from foreign countries" (quoted in Winarto 2005, 2).

It is interesting to see how two very different concepts have been proposed to rectify the perceived difficulties. Both food security and food sovereignty can be considered as a response to the Green Revolution's environmental, nutritional, and social externalities. But, as Foucault pointed out, a particular interpretation of a "problem" can invite different responses. Diagnosis and prescription are very different in food security and food sovereignty.

3. http://www.poptel.org.uk/panap/latest/wfs3.htm.

References

Abraham, Itty. 1998. *The Making of the Indian Atomic Bomb: Science, Secrecy and the Postcolonial State.* London: Zed Books.

———. 2000. "Postcolonial Science, Big Science, and Landscape." In *Doing Science and Culture,* ed. R. Reid and S. Traweek, 49–70. London: Routledge.

Achmad, Januar. 1999. *Hollow Development: The Politics of Health in Soeharto's Indonesia.* Canberra: Australian National University.

Aditjondro, George. 1998. "Suharto Inc," April [accessed 7 July 2005]. http://www.malra.org/posko/malra.php4?nr = 15074.

———. 2000. "Chopping the Global Tentacles of the Suharto Oligarchy" [accessed 8 July 2004]. http://www.unhas.ac.id/~rhiza/gja1.html.

Agrawal, Arun. 2005. *Environmentality: Technologies of Government and the Making of Subjects.* Durham, NC: Duke University Press.

Alderman, H., J. Hoddinott, and B. Kinsey. 2006. "Long-Term Consequences of Early Childhood Malnutrition." *Oxford Economic Papers* 58, no. 3: 450–74.

Allen, Lindsay H. 2003. "Interventions for Micronutrient Deficiency Control in Developing Countries: Past, Present and Future." *Journal of Nutrition* 133, no. 11: 3875–78S.

Allen, Patricia. 2004. *Together at the Table: Sustainability and Sustenance in the American Agrifood System.* Rural Studies Series. University Park: Pennsylvania State University Press with the Rural Sociological Society.

Allen, Patricia, Margaret FitzSimmons, Michael Goodman, and Keith Warner. 2003. "Shifting Plates in the Agrifood Landscape: The Tectonics of Alternative Agrifood Initiatives in California." *Journal of Rural Studies* 19:61–75.

Allen, Patricia, and Carolyn Sachs. 2007. "Women and Food Chains: The Gendered Politics of Food." *International Journal of Sociology of Food and Agriculture* 15, no. 1: 1–23.

Alston, J. M., S. Dehmer, and P. G. Pardey. 2006. "International Initiatives in Agricultural R & D: The Changing Fortunes of the CGIAR." In *Agricultural R & D in the Developing World: Too Little, Too Late?,* ed. Philip G. Pardey, Julian M. Alston, and Roley R. Piggott, 313–60. Washington, DC: IFPRI.

Anderson, Kym, and L. Alan Winters. 2008. "The Challenge of Reducing International Trade and Migration Barriers." Policy Research Working Paper. Washington, DC: World Bank.

Anderson, Warwick. 2002. "Going through the Emotions: American Public Health and Colonial Mimicry." *American Literary History* 14, no. 4: 686–719.

Appadurai, Arjun. 1988. *The Social Life of Things: Commodities in Cultural Perspective.* Cambridge: Cambridge University Press.

Apple, Rima D. 1987. *Mothers and Medicine: A Social History of Infant Feeding, 1890–1950.* Madison: University of Wisconsin Press.

———. 1995. "Constructing Mothers: Scientific Motherhood in the Nineteenth and Twentieth Centuries." *Social History of Medicine* 8, no. 2: 161–78.

———. 1996. *Vitamania: Vitamins in American Culture.* Health and Medicine in American Society. New Brunswick, NJ: Rutgers University Press.

——. 1997. "Science Gendered: Nutrition in the United States, 1840–1940." *Kappa Omicron Nu FORUM* 10, no. 1: 11–34.

——. 2006. *Perfect Motherhood: Science and Childrearing in America.* New Brunswick, NJ: Rutgers University Press.

APTINDO (Indonesian Association of Wheat Flour Producers). 2001. "SNI Terigu Segera Diberlakukan." 24 February. Jakarta: APTINDO.

——. 2003. "Laporan." Jakarta: APTINDO.

Arifin, Bustanil. 1993. *Pangan Dalam Orde Baru.* Jakarta: Koperasi Jasa Informasi.

Arnelia, and Sri Muljati. 1993. "Praktek Pemberian Makanan Pada Bayi Di Bogor Dan Faktor-Faktor Sosial Budaya Yang Mempengaruhi." *Penelitian Gizi Dan Makanan* 16: 29–37.

Arnold, David. 1993. *Colonizing the Body: State Medicine and Epidemic Disease in Nineteenth-Century India.* Berkeley: University of California Press.

Aronson, Naomi. 1982. "Nutrition as a Social Problem: A Case Study of Entrepreneurial Strategy in Science." *Social Problems* 29, no. 5: 474–87.

Asian Development Bank (ADB). 2000a. *Annual Report 1999.* Manila: ADB.

——. 2000b. *Manila Forum 2000: Strategies to Fortify Essential Foods in Asia and the Pacific.* Nutrition and Development Series. Manila: ADB.

——. 2000c. "Technical Assistance: Regional Initiative to Eliminate Micronutrient Deficiency in Asia through Public-Private Partnership." Manila: ADB.

——. 2001. *Investing in Child Nutrition in Asia.* Manila: ADB.

Atmarita. 2005. "Nutrition Problems in Indonesia" [accessed 5 August 2007]. http://www.gizi.net/.download/nutrition%20problem%20in%20Indonesia.pdfS.

Avakian, Arlene Voski, and Barbara Haber. 2005. *From Betty Crocker to Feminist Food Studies: Critical Perspectives on Women and Food.* Amherst: University of Massachusetts Press.

Azwar, A. 2004. "Kecenderungan Masalah Gizi Dan Tantangan Di Masa Datang." Paper presented at the *Pertemuan Advokasi Program Perbaikan Gizi Menuju Keluarga Sadar Gizi,* Hotel Sahid, Jakarta, Indonesia. September 27.

Backstrand, J. R. 2002. "The History and Future of Food Fortification in the United States." *Nutrition Reviews* 60, no. 1: 15–26.

Badan Kerja Peningkatan Penggunaan ASI and Yayasan ASI Indonesia (BKPP-ASI and YASIA). 2003. "Taktik Baru Pemasaran Dan Pelanggaran Kode Internasional Pemasaran Pasi Di Indonesia 2003." Jakarta: BKPP-ASI and YASIA.

Badinter, Elisabeth. 1981. *The Myth of Motherhood: An Historical View of the Maternal Instinct.* London: Souvenir Press.

Bagla, Pallava. 1998. "Midlife Crisis Threatens Center for Semiarid Crops." *Science* 279, no. 5347: 26–27.

Barndt, Deborah. 2002. *Tangled Routes: Women, Work, and Globalization on the Tomato Trail.* Lanham, MD: Rowman & Littlefield.

Barnes, B. 1977. *Interests and the Growth of Knowledge.* London: Routledge.

Barrientos, Stephanie. 1997. "The Hidden Ingredient: Female Labour in Chilean Fruit Exports." *Bulletin of Latin American Research* 16, no. 1: 71–81.

Baru, Ramu, and Amar Jesani. 2000. "The Role of the World Bank in International Health: Renewed Commitment and Partnership." *Social Science and Medicine* 50:183–84.

Basta, Samir S., Soekirman, Darwin Karyadi, and Nevin S. Scrimshaw. 1979. "Iron Deficiency Anemia and the Productivity of Adult Males in Indonesia." *American Journal of Clinical Nutrition* 32:916–25.

Baumslag, Naomi, and Dia L. Michels. 1995. *Milk, Money, and Madness: The Culture and Politics of Breastfeeding.* Westport, CT: Bergin and Garvey.

BBC News. 2005. "Monsanto Fined $1.5m for Bribery." 7 January.

Beall, J. 1997. "'In Sickness and in Health': Gender Issues in Health Policy and Their Implications for Development in the 1990s." In *Searching for Security: Women's Responses to Economic Transformation,* ed. I. Baud and I. Smyth, 67–95. London: Routledge.

Beardsworth, Alan, and Teresa Keil. 1997. *Sociology on the Menu: An Invitation to the Study of Food and Society.* London: Routeledge.

Beaton, Gh. H., R. Martorell, K. J. Aronson, B. Edmonston, G. McCabe, A. C. Ross, and B. Harvey. 1992. *Effectiveness of Vitamin A Supplementation in the Control of Young Child Morbidity and Mortality in Developing Countries.* Geneva: UN ACC/SCN.

Belair, Felix, Jr. 1965. "Hunger Imperils US Aid Program." *New York Times,* 18 July.

Belasco, Warren. 1993. *Appetite for Change: How the Counterculture Took on the Food Industry.* Ithaca: Cornell University Press.

———. 2006. *Meals to Come: A History of the Future of Food.* Berkeley: University of California Press.

Belavady, Bhavani, and C. Gopalan. 1959. "Chemical Composition of Human Milk in Poor Indian Women." *Indian Journal of Medical Research* 47, no. 2: 234–45.

Bentley, Amy. 1998. *Eating for Victory: Food Rationing and the Politics of Domesticity.* Urbana-Champaign: University of Illinois Press.

Berg, Alan, David L. Call, and Nevin S. Scrimshaw. 1973. *Nutrition, National Development, and Planning: Proceedings of an International Conference.* Cambridge: MIT Press.

Berg, Alan, and Robert J. Muscat. 1973. *The Nutrition Factor: Its Role in National Development.* Washington, DC: Brookings Institution.

Bidan Litbang Pertanian. 2004. "Rencana Strategis Badan Penelitian Dan Pengemban-gan Pertanian 2005–2009." Jakarta: Bidan Litbang Pertanian.

Bill and Melinda Gates Foundation. 2011. "Agricultural Development Strategy Overview 2011" [accessed 2 February 2012]. http://www.gatesfoundation.org/agricultural development/Documents/agricultural-development-strategy-overview.pdf.

Binimelis, R., W. Pengue, and I. Monterroso. 2009. "Transgenic Treadmill: Responses to the Emergence and Spread of Glyphosate-Resistant Johnsongrass in Argentina." *Geoforum* 40, no. 4: 623–33.

Binswanger, Hans P., and Pierre Landell-Mills. 1995. *The World Bank's Strategy for Reducing Poverty and Hunger: A Report to the Development Community.* Environmentally Sustainable Development Studies and Monographs Series, no. 4. Washington, DC: World Bank.

Bishai, David, and Ritu Nalubola. 2002. "The History of Food Fortification in the United States: Its Relevance for Current Fortification Efforts in Developing Countries." *Economic Development and Cultural Change* 51, no. 1: 37–53.

Bogasari Flour Mills. 1996."25 Years of Helping to Feed a Nation." Corporate PR brochure. Jakarta: Bogasari Flour Mills.

Bonanno, Alessandro, Lawrence Busch, William Friedland, and Lourdes Guiveia. 1994. *From Columbus to ConAgra: The Globalization of Agriculture and Food.* Lawrence: University Press of Kansas.

Bonneuil, Christophe. 2001. "Development as Experiment: Science and State Building in Late Colonial and Postcolonial Africa, 1930–1970." *Osiris* 15:258–81.

Bouis, Howarth. 1994. *Agricultural Technology and Food Policy to Combat Iron Deficiency in Developing Countries.* Washington, DC: IFPRI.

———. 2003. "Micronutrient Fortification of Plants through Plant Breeding: Can It Improve Nutrition in Man at Low Cost?" *Proceedings of the Nutrition Society* 62, no. 2: 403–11.

Bouis, Howarth, Robin D. Graham, and Ross M. Welch. 1999. *The CGIAR Micro-nutrients Project: Justification, History, Objectives, and Summary of Findings.* Washington, DC: IFPRI.

Brantley, Cynthia. 1997. "Kikuyu-Maasai Nutrition and Colonial Science: The Orr and Gilks Study in Late 1920s Kenya Revisited." *International Journal of African Historical Studies* 30, no. 1: 49–86.

Bressani, Ricardo. 1984. "Incorporating Nutritional Concerns into the Specification of Desired Changes in Commodity Characteristics in International Agricultural Research." In *International Agricultural Research and Human Nutrition,* ed. Per Pinstrup-Andersen, Alan Berg, and Martin Forman, 245–64. Washington, DC: IFPRI.

Brooks, H., and R. B. Johnson. 1991. "Comments: Public Policy Issues." In *The Genetic Revolution: Scientific Prospects and Public Perceptions,* ed. B. Davies. Baltimore: Johns Hopkins University Press.

Brooks, Sally. 2005. "Biotechnology and the Politics of Truth: From the Green Revolution to an Evergreen Revolution." *Sociologia Ruralis* 45, no. 4: 360–79.

——. 2010. *Rice Biofortification: Lessons for Global Science and Development.* London: Earthscan.

——. 2011. "Is International Agricultural Research a Global Public Good?: The Case of Rice Biofortification." *Journal of Peasant Studies* 38, no. 1: 67–80.

Brown, David. 1994. "It's Cheap and Effective, with Wonders Still Being (Re)Discovered." *Washington Post,* 7 November.

Brown, James P. 1968. "Food Science Narrows the 'Edibility Gap.'" *New York Times,* 25 November.

Brown, K., K. Dewey, and L. Allen. 1998. "Proteins and Micronutrients Required from Complementary Foods." *Complementary Feeding of Young Children in Developing Countries: A Review of Current Scientific Knowledge,* WHO/NUT/98.1. Geneva, Switzerland: WHO.

Brown, Phil, and F. T. Ferguson. 1995. "Making a Big Stink: Women's Work, Women's Relationships, and Toxic Waste Activism." *Gender & Society* 9:145–67.

Brown, Phil, S. McCormick, Brian Mayer, Stephen Zavestoski, Rachel Morello-Forsch, Rebecca Gasior Altman, and Laura Senier. 2006. "'A Lab of Our Own': Environmental Causation of Breast Cancer and Challenges to the Dominant Epidemiological Paradigm." *Science, Technology, and Human Values* 31, no. 5: 499–536.

Brown, Wendy. 1995. *States of Injury: Power and Freedom in Late Modernity.* Princeton: Princeton University Press.

Burnet, E., and W. R. Aykroyd. 1935. "Nutrition and Public Health." *Quarterly Bulletin of the Health Organizations of the League of Nations* 4:1–140.

Burnett, John. 1979. *Plenty and Want: A Social History of Diet in England from 1815 to the Present Day.* London: Scolar Press.

Busch, Lawrence. 2011. *Standards: Recipes for Reality.* Cambridge: MIT Press.

Business Alliance for Food Fortification (BAFF). 2005. "The First Annual Meeting of Business Alliance for Food Fortification," October [accessed 16 May 2005]. http://www.gainhealth.org/gain/ch/en-en/file.cfm/baff_report2.pdf?contentID = 1507.

Butler, Judith. 2000. "Competing Universalities." In *Contingency, Hegemony, Universality: Contemporary Dialogues on the Left,* ed. J. Butler, E. Laclau, and S. Žižek. London. Verso.

Buttel, Frederick H. 2003. "The Global Politics of GEOs: The Achilles' Heel of the Globalization Regime?" In *Engineering Trouble: Biotechnology and Its Discontents,* ed. Rachel A. Schurman, Dennis Doyle, and Takahashi Kelso, 152–73. Berkeley: University of California Press.

Byerlee, D., and H. Dubin. 2009. "Crop Improvement in the CGIAR as a Global Success Story of Open Access and International Collaboration." *International Journal of the Commons* 4, no. 1: 452–80.

Calabro, Karen S., Karie A. Bright, and Saroj Bahl. 2001. "International Perspectives: The Profession of Dietetics." *Nutrition* 17:594–99.

Caldwell, John, and Pat Caldwell. 1986. *Limiting Population Growth and the Ford Foundation Contribution.* Dover, NH: Frances Pinter.

Callon, M., P. Lascoumes, and Y. Barthe. 2009. *Acting in an Uncertain World: An Essay on Technical Democracy.* Cambridge: MIT Press.

Campbell, Hugh, and Jane Dixon. 2009. "Introduction to the Special Symposium: Reflecting on Twenty Years of the Food Regimes Approach in Agri-Food Studies." *Agriculture and Human Values* 26, no. 4: 261–65.

Canadian International Development Agency (CIDA). 2006. "*Report on Indonesian-UNICEF Iodine Deficiency Disorder Control Project*" [accessed 13 May 2012]. http://www.acdi-cida.gc.ca/acdi-cida/acdi-cida.nsf/eng/EMA-218132549-PNM.

Cannon, Geoffrey. 2002. "Nutrition: The New World Map." *Asia Pacific Journal of Clinical Nutrition* 11:S480–97.

Carney, Judith A. 1994. "Contracting a Food Staple in the Gambia." In *Living under Contrac,* ed. Peter D. Little and Michael Watts, 167–87. Madison: University of Wisconsin Press.

Carpenter, Kenneth J. 1994. *Protein and Energy: A Study of Changing Ideas in Nutrition.* Cambridge: Cambridge University Press.

——. 2003a. "A Short History of Nutritional Science: Part 1 (1785–1885)." *Journal of Nutrition* 133:638–45.

——. 2003b. "A Short History of Nutritional Science: Part 2 (1885–1912)." *Journal of Nutrition* 133:975–84.

——. 2003c. "A Short History of Nutritional Science: Part 3 (1912–1944)." *Journal of Nutrition* 133:3023–32.

——. 2003d. "A Short History of Nutritional Science: Part 4 (1945–1985)." *Journal of Nutrition* 133:3331–42.

Carr, Marilyn, Martha Alter Chen, and Jane Tate. 2000. "Globalization and Home-Based Workers." *Feminist Economics* 6, no. 3: 123–42.

Carroll, Rory. 2002. "GM Firms the Only Winners at Food Talks Summit." *Guardian.* June 13.

Chase, Susan E., and Mary F. Rogers. 2001. *Mothers and Children: Feminist Analyses and Personal Narratives.* New Brunswick, NJ: Rutgers University Press.

Chen, Lincoln C. 1986. "Nutrition in Developing Countries and the Role of the International Agencies: In Search of a Vision." In *Nutrition Issues in Developing Countries for the 1980s and 1990s: Proceedings of a Symposium,* 63–82. Washington, DC: National Academy Press.

Chong, M. 2003. "Acceptance of Golden Rice in the Phillippines 'Rice Bowl.'" *Nature Biotechnology* 21, no. 9: 971–72.

Clapp, Jennifer. 2005. "The Political Economy of Food Aid in an Era of Agricultural Biotechnology." *Global Governance* 11, no. 4: 467-85.

Cleaver, Harry M., Jr. 1972. "The Contradictions of the Green Revolution." *American Economic Review* 62, no. 2: 177–86.

Clifton, Deborah, and Fiona Gell. 2002. "Saving and Protecting Lives by Empowering Women." In *Gender, Development, and Humanitarian Work,* ed. C. Weetman. London: Oxfam.

Colfer, Carol J. Pierce. 1991. "Indigenous Rice Production and the Subtleties of Culture Change: An Example from Borneo." *Agriculture and Human Values* 8, no. 1–2: 67–84.

Collier, Paul. 2008. "Politics of Hunger: How Illusion and Greed Fan the Food Crisis" *Foreign Affairs* 87: 67–79.

Collins, Jane. 1993. "Gender, Contracts and Wage Work: Agricultural Restructuring in Brazil's Sao Francisco Valley." *Development and Change* 24: 53-82.

——. 1995. "Gender and Cheap Labor in Agriculture." *Food and Agrarian Orders in the World-Economy,* ed. Philip McMichael, 217–32. Westport, CT: Greenwood Press.

Collins, Patricia Hill. 1998. "It's All in the Family: Intersections of Gender, Race, and Nation." *Hypatia* 13, no. 3: 62–82.

——. 1999. "Producing the Mothers of the Nation: Race, Class, and Contemporary US Population Policies." In *Women, Citizenship and Difference,* ed. Nira Yuval-Davis and Pnina Werbner, 118–29. London: Zed Books.

Comey, John, and David Liebhold. 1999. "The Family Firm." *Time,* 24 May.

Consumers Association of Penang. 1986. *Selling Dreams: How Advertising Misleads Us.* Penang, Malaysia: Consumers Association of Penang.

Consumer International. 2008. *Junk Food Trap: Marketing Unhealthy Food to Children in Asia Pacific.* London: Consumer International.

Conway, Gordon. 1998. *The Doubly Green Revolution: Food for All in the Twenty-First Century.* Ithaca: Cornell University Press.

Cornia, Govanni, Richard Jolly, and Francis Stewart. 1987. *Adjustment with a Human Face.* Oxford: Oxford University Press.

Crouch, Harold. 2007. *The Army and Politics in Indonesia.* Jakarta: Equinox.

Cullather, Nick. 2004. "Miracles of Modernization: The Green Revolution and the Apotheosis of Technology." *Diplomatic History* 28, no. 2: 227–54.

Dahro, A. M., Dewi Permaesih, Muhilal, Darwin Karyadi, and Bambang Setiohadji. 1991. "Masalah Anemia Di Empat Provinsi Wilaya Indonesia Bagian Timur." *Gizi Indonesia* 16, no. 1–2: 9–14.

Dan-Cohen, T., and P. Rabinow. 2006. *A Machine to Make a Future: Biotech Chronicles.* Princeton: Princeton University Press.

Darnton-Hill, Ian, Jose O. Mora, Herbert Weinstein, Steven Wilbur, and P. R. Nalubola. 1999. "Iron and Folate Fortification in the Americas to Prevent and Control Micronutrient Malnutrition: An Analysis." *Nutrition Reviews* 57, no. 1: 25–31.

Darnton-Hill, Ian, and R. Nalubola. 2002. "Fortification Strategies to Meet Micronutrient Needs: Successes and Failures." *Proceedings of the Nutrition Society* 61, no. 2: 231–41.

Datta, K., N. Baisakh, N. Oliva, L. Torrizo, E. Abrigo, J. Tan, M. Rai, S. Rehana, S. Al-Babili, P. Beyer, I. Potrykus, and S. Datta. 2003. "Bioengineered 'Golden' Indica Rice Cultivars with Beta Carotene Metabolism in the Endosperm with Hygromycin and Mannose Selection Systems." *Plant Biotechnology Journal* 1, no. 2: 81–90.

Dawe, D., R. Robertson, and L. Unnevehr. 2002. "Golden Rice: What Role Could It Play in Alleviation of Vitamin A Deficiency?" *Food Policy* 27, no. 5–6: 541–60.

Deacon, Roger. 2000. "Theory as Practice: Foucault's Concept of Problematization." *Telos* 118: 127–33.

de Pee, Saskia, Martin W. Bloem, Yip Ray Satoto, Asmira Sukaton, Roy Tjiong, Roger Muhilal Shrimpton, and Benny Kodyat. 1998. "Impact of a Social Marketing Campaign Promoting Dark-Green Leafy Vegetables and Eggs in Central Java, Indonesia." *International Journal for Vitamin and Nutrition Research* 68, no. 6: 389–98.

de Pee, Saskia, J. Diekhans, G. Stallkamp, L. Kiess, Regina Moench-Pfanner, E. Martini, M. Sari, A. Stormer, S. Kosen, and M. W. Bloem. 2002. "Breastfeeding and Complementary Feeding Practices in Indonesia: Nutrition and Health Surveillance Report." Jakarta: Helen Keller Worldwide.

DeLind, Laura, and Anne Ferguson. 1999. "Is This a Woman's Movement?: The Relationship of Gender to Community-Supported Agriculture in Michigan." *Human Organization* 58:190–200.

Desmarais, Annette Aurelie. 2004. "The Via Campesina: Peasant Women on the Frontier of Food Sovereignty." *Canadian Woman Studies* 23, no. 1: 140–45.

———. 2007. *La Vía Campesina: Globalization and the Power of Peasants.* Halifax: Fernwood.

de Waal, Alex. (1989) 2005. *Famine That Kills: Darfur, Sudan.* Oxford: Oxford University Press.

Dewey, K. G. 1979. "Agricultural Development, Diet, and Nutrition." *Ecology of Food and Nutrition* 8:265–73.

Dibb, Sally. 1999. "The Impact of the Changing Marketing Environment in the Pacific Rim: Four Case Studies." *International Journal of Retail & Distribution Management* 24, no. 11: 16-29.

Direktorat Bina Gizi Masyarakat. Departmen Kesehatan. 1992."Laporan: Program Perbaikan Gizi Tahun 1991/1992." Jakarta: Department Kesehatan.

———. 1994."Laporan Pelaksanaan Proyek Perbaikan Gizi Tahun 1993/1994." Jakarta: Department Kesehatan.

———. 1996. "Laporan: Program Perbaikan Gizi Tahun 1995/1996." Jakarta: Department Kesehatan.

———. 1997. "Laporan Tahunan Proyek Perbaikan Gizi Tahun 1996/1997." Jakarta: Department Kesehatan.

———. 2001. "Laporan Tahunan Proyek Perbaikan Gizi Tahun 2001." Jakarta: Department Kesehatan.

Dixon, Jane. 2002. *The Changing Chicken: Chooks, Cooks and Culinary Culture.* Sydney: University of New South Wales Press, 2002.

Dixon, Jane, and Cathy Banwell. 2004. "Re-embedding Trust: Unravelling the Construction of Modern Diets." *Critical Public Health* 14, no. 2: 117–31.

Dolan, Catherine. 2001. "The 'Good Wife': Struggles over Resources in the Kenyan Horticultural Sector." *Journal of Development Studies* 37, no. 3: 39–70.

Douglas, Mary. 1966. *Purity and Danger.* London: Routledge and Kegan Paul.

Douglas, Mary, Des Gasper, Michael Thompson, and Steven Ney. 1998. "Human Needs and Wants." In *Human Choice and Global Change.*Vol. 1, *The Societal Framework,* ed. Steve Rayner and Elizabeth Malone, 195–263. Columbus, OH: Batelle Press.

Dunn, Elizabeth C. 2004. *Privatizing Poland: Baby Food, Big Business, and the Remaking of Labor.* Ithaca: Cornell University Press.

Dunne, Nancy. 1994. "World Bank Urged to Set Up Nutrition Programme." *Financial Times,* 17 December.

DuPuis, E. Melanie. 2002. *Nature's Perfect Food: How Milk Became America's Drink.* New York: New York University Press.

Eckholm, Erik. 1985. "Diet Deficiency of Vitamin A Is Revealed as a Major Killer." *New York Times,* 3 September.

Economist. 2004. "Food for Thought." *Economist* 372:71.

Edmunds, Lavinia. 1989. "The Magic Bullet?" *Johns Hopkins Magazine,* 13–20.

Ehrenreich, Barbara, and Deirdre English. 1978. *For Her Own Good: 150 Years of the Experts' Advice to Women.* Garden City, NY: Anchor Books.

Ehrlich, Paul R. 1968. *The Population Bomb.* New York: Ballantine Books.

Enarson, E., and L. Meyreles. 2004. "International Perspectives on Gender and Disaster: Differences and Possibilities." *International Journal of Sociology and Social Policy* 24, no. 10: 49–93.

Epstein, Steven. 2000. "Democracy, Expertise, and AIDS Treatment Activism." In *Science, Technology, and Democracy,* ed. Daniel Lee Kleinman, 15–32. New York: State University of New York Press.

Escobar, Arturo. 1984. "Discourse and Power in Development: Michel Foucault and the Relevance of His Work to the Third World." *Alternatives* 10: 77–400.

——. 1995. *Encountering Development: The Making and Unmaking of the Third World.* Princeton Studies in Culture/Power/History. Princeton: Princeton University Press.

——. 1996. "Construction Nature: Elements for a Post-Structuralist Political Ecology." *Futures,* 28, no. 4: 325–43.

Eviandaru, M., D. S. Indriaswati, Rika Pratiwi, Sri Sulistyani, R. A. Wigati, Arimbi, and Karen Washburn, eds. 2001. *Perempuan Postkolonial Dan Identitas Komoditi Global.* Yogyakarta, Indonesia: Lembaga Studi Realino.

Fabiosa, Jacinto F. 2006. *Westernization of the Asian Diet: The Case of Rising Wheat Consumption in Indonesia.* Center for Agricultural and Rural Development, Ames [accessed 19 July 2011]. http://www2.econ.iastate.edu/research/webpapers/paper_12587.pdf.

Fair, Mollie. 2008. *From Population Lending to HNP Results: The Evolution of the World Bank's Strategies in Health, Nutrition, and Population: Background Paper for the IEG Evaluation of World Bank Support for Health, Nutrition, and Population."* Washington, DC: World Bank.

Feldbaum, Carl B. 2002. "Biotechnology's Foreign Policy" [accessed 16 May 2001]. www.bio.org.

Ferguson, James. 1990. *The Anti-Politics Machine: "Development," Depoliticization, and Bureaucratic Power in Lesotho.* Cambridge: Cambridge University Press.

Ferguson, James, and Akhil Gupta. 2002. "Spatializing States: Toward an Ethnography of Neoliberal Governmentality." *American Ethnologist* 29:981–1002.

Field, John Osgood. 1987. "Multisectoral Nutrition Planning: A Post-Mortem." *Food Policy* 12, no. 1: 15–28.

Fleuret, Patrick, and Anne Fleuret. 1980. "Nutrition, Consumption, and Agricultural Change." *Human Organization* 39:250–60.

Flinn, J. C., and L. J. Unnevehr. 1984. "Contributions of Modern Rice Varieties to Nutrition in Asia: An IRRI Perspective." In *International Agricultural Research and Human Nutrition,* ed. Per Pinstrup-Andersen, Alan Berg, and Martin Forman, 157–78. Washington, DC: IFPRI.

Fointuna, Y., and Agus Maryono. 2009. "Malnutrition Claims 6 Lives in Kupang." *Jakarta Post,* 9 January.

Food and Agriculture Organization (FAO). 1962. *Freedom from Hunger Campaign.* Basic Study no. 5. Rome: FAO.

——. 1996. *Rome Declaration on World Food Security.* Rome: FAO

——. 2002a. *Report of the World Food Summit: Five Years Later (Part 1).* Rome: FAO.

——. 2002b. *State of the Food Insecurity in the World.* Rome: FAO.

——. 2011. *The State of Food and Agriculture Report,* 2010–2011. Rome: FAO.

FAO, and WHO. 1992. *International Conference on Nutrition, World Declaration and Plan of Action* [accessed 13 May 2012]. http://www.fao.org/docrep/U9920t/u9920t0a.htm.

Foucault, Michel.1980. *Power/Knowledge: Selected Interviews and Other Writings.* New York: Pantheon Press.

——. 1981. "The Order of Discourse." In *Untying the Text: A Post-Structuralist Reader,* ed. Robert Young, 51–78. Boston: Routledge and Kegan Paul.

——. 1991. "Questions of Method." In *The Foucault Effect,* ed. Graham Burchell, Colin Gordon, and Peter Miller, 53–72. Chicago: University of Chicago Press.

Foucault, Michel, and Paul Rabinow. 1984. *Foucault Reader.* New York: Pantheon.

Frank, D.1985. "Housewives, Socialists, and the Politics of Food: The 1917 New York Cost-of-Living Protests." *Feminist Studies* 11:255–86.

Fraser, Nancy. 1989. *Unruly Practices: Power, Discourse and Gender in Contemporary Social Theory.* Cambridge: Polity.

———. 1997. *Justice Interruptus: Critical Reflections of the 'Postsocialist' Condition.* London: Routledge.

———. 2009. *Scales of Justice: Reimagining Political Space in a Globalizing World.* New York: Columbia University Press.

Friedmann, Harriet. 1982. "The Political Economy of Food: The Rise and Fall of the Postwar International Food Order." *American Journal of Sociology* S88: S248–86.

———.1987. "Family Farms and International Food Regimes." In *Peasants and Peasant Societies,* ed. T. Shanin, 247–58. Oxford: Basil Blackwell.

———.1991. "Changes in the International Division of Labor: Agri-Food Complexes and Export Agriculture." In *Towards a New Political Economy of Agriculture,* ed. William H. Friedland, 65–93. Boulder, CO: Westview Press.

———. 1993. "The Political Economy of Food: A Global Crisis." *New Left Review:* 29–57.

———. 1999. "Remaking 'Traditions': How We Eat, What We Eat and the Changing Political Economy of Food." In *Women Working the NAFTA Food Chain: Women Food and Globalization,* ed. Deborah Barndt, 36–60. Toronto: Second Story Press.

———. 2005. "From Colonialism to Green Capitalism: Social Movements and Emergence of Food Regimes." In *New Directions in the Sociology of Global Development: Research in Rural Sociology and Development* 11:227–64.

Friedmann, Harriet, and P. McMichael. 1989. "Agriculture and the State System: The Rise and Decline of National Agricultures, 1870 to the Present." *Sociologia Ruralis* 29:93–117.

Friend, Theodore. 2003. *Indonesian Destinies.* Cambridge: Belknap Press of Harvard University Press.

Fujimura, Joan H. 1992. "Crafting Science: Standardized Packages, Boundary Obejcts, and 'Translation.'" In *Science as Practice and Culture,* ed. A. Pickering. Chicago: University of Chicago Press.

Fulu, Emma. 2007. "Gender, Vulnerability, and the Experts: Responding to the Maldives Tsunami." *Development and Change* 38, no. 1: 843–64.

GAIN (Global Alliance for Improved Nutrition). 2005. "The Business Alliance for Food Fortification (BAFF)" [accessed 16 May 2006]. http://www.gainhealth.org/gain/ch/en-en/index.cfm?page = /gain/home/activities/business_consumer_programs/business_action_network.

GAIN, and BAFF. 2005. "GAIN Business Alliance" [accessed 19 May 2006]. http://www.gainhealth.org/gain/ch/en-en/index.cfm?page = /gain/home/activities/business_consumer_programs/business_action_network.

Gardner, Gary, and Brian Halweil. 2000. *Underfed and Overfed: The Global Epidemic of Malnutrition.* Washington, DC: World Watch Institute.

Gardner, Katy, and David Lewis. 1996. *Anthropology, Development, and the Post-Modern Challenge.* London: Pluto Press.

Gaskell, G. 2000. "Agricultural Biotechnology and Public Attitudes in the European Union." *AgBio Forum* 3, no. 2–3: 87–96.

GATRA. 2005. "Kurang Gizi di Lumbung Padi." *GATRA,* 7 July.

Gavaghan, Colin. 2009. "'You Can't Handle the Truth': Medical Paternalism and Prenatal Alcohol Use." *Journal of Medical Ethics* 35: 300–303.

Ghosh, Jayati. 2010. "The Unnatural Coupling: Food and Global Finance." *Journal of Agrarian Change* 10, no. 1: 72–86.

Gieryn, T. 1983. "Boundary Work and the Demarcation of Science from Non-Science: Strains and Interests in Professional Ideologies of Scientists." *American Sociological Review* 48: 781–95.

Gilks, J. L., and J. B. Orr. 1927. "The Nutritional Condition of the East African Native." *Lancet* 29, no. 5402: 560–62.

Gillespie, S., and J. Mason. 1994. "Controlling Vitamin A Deficiency." ACC/SCN Nutrition Policy Discussion Paper, no. 1. Geneva: UN ACC/SCN.

Golden Rice Humanitarian Board. 2005. "Statement from the Golden Rice Humanitarian Board" [accessed 2 July 2005]. http://www.agbioworld.org/newsletter_wm/index.php?caseid = archive&newsid = 2340.

Goldman, M. 2001. "Constructing an Environmental State: Eco-Governmentality and Other Transnational Practices of a 'Green' World Bank." *Social Problems* 48, no. 4: 499–523.

Goodman, David, and M. Goodman. 2001. "Sustaining Foods: Organic Consumption and the Socio-Ecological Imaginary." In *Exploring Sustainable Consumption: Environmental Policy and the Social Sciences,* ed. M. Cohen, and J. Murphy, 97–119. Oxford: Elsevier Science.

Goodman, David, and M. R. Redclift. 1991. *Refashioning Nature: Food, Ecology, and Culture.* London: Routledge.

Government of Australia. Department of Foreign Affairs and Trade. 2000. *"Indonesia: Facing the Challenge"* [accessed 18 July 2005]. http://www.dfat.gov.au/publications/indonesia/index.html.

Government of Indonesia. 1993. *Rencana Pembangunan Lima Tahun VI (Repelita VI).* Jakarta: Government of Indonesia.

——. 2001. *The National Household Survey.* Jakarta: Government of Indonesia.

Gozal, Edwin. 1998. "Who Will Benefit from the IMF Bailout?" [accessed 5 July 2005]. http://www.xs4all.nl/~peace/pubeng/mov/movto/ipeg.html.

Graham, R. D. "Biofortification: a Global Challenge Program." *International Rice Research Notes* 29, no. 1 (2003): 4–8.

Graham, Wendy. 2002. "Now or Never: The Case for Measuring Maternal Mortality." *Lancet* 359, no. 9307: 701–4.

GRAIN (Genetic Resources Action International). 2009. "Twelve Years of GM Soya in Argentina: A Disaster for People and the Environment." *Seedling* [accessed 5 May 2011]. http://www.grain.org/article/entries/706-twelve-years-of-gm-soya-in-argentina-a-disaster-for-people-and-the-environment.

Grant, Julia. 1998. *Raising Baby by the Book: The Education of American Mothers.* New Haven: Yale University Press.

Grossman, Lawrence S. 1998. *The Political Ecology of Bananas: Contract Farming, Peasants, and Agrarian Change in the Eastern Caribbean.* Chapel Hill: University of North Carolina Press.

Gugliotta, Guy. 2000. "Gene-Altered Rice May Help Fight Vitamin A Deficiency Globally." *Washington Post,* 14 January.

Gupta, Akhil. 1998. *Postcolonial Developments: Agriculture in the Making of Modern India.* Durham, NC: Duke University Press.

——. 2001. "Governing Population: The Integrated Child Development Services in India." In *States of Imagination,* ed. T. Blom Hansen and F. Stepputat. Durham, NC: Duke University Press.

Gura, T. 1999. "New Genes Boost Rice Nutrients." *Science* 285, no. 5430: 994.

Guthman, Julie. 2007. "The Polanyian Way?: Voluntary Food Labels as Neoliberal Governance." *Antipode* 39:456–78.

Hacking, Ian. 1989. *The Taming of Chance.* Cambridge: Cambridge University Press.

Haddad, Lawrence. 1999. "Women's Status: Levels, Determinants, Consequences for Malnutrition, Interventions, and Policy." *Asia Development Review* 17, no. 1–2: 96–131.

———. 2003. "Redirecting the Diet Transition: What Can Food Policy Do?" *Development Policy Review* 21, no. 5–6: 599–614.

Hadiz, Uedi. 2000. "Globalization, Labour and the State: The Case of Indonesia." *Asia Pacific Business Review* 6, no. 3–4: 239—59.

Haney, Lynne. 2002. *Inventing the Needy Gender and the Politics of Welfare in Hungary.* Berkeley: University of California Press.

Hansen, Gary E. 1978. "Bureaucratic Linkages and Policy-Making in Indonesia: BIMAS Revisited." In *Political Power and Communications in Indonesia,* ed. Karl D. Jackson and Lucian W. Pye, 322–42. Berkeley: University of California Press.

Hansen-Kuhn, Karen. 2011. "Making US Policy Serve Global Food Security Goals" [accessed 3 December 2012]. http://www.iatp.org/files/451_2_107901.pdf.

Hartini, S., A. Winkvist, L. Lindholm, H. Stenlund, V. Persson, D. S. Nurdiati, and A. Surjono. 2003. "Nutrient Intake and Iron Status of Urban Poor and Rural Poor without Access to Rice Fields Are Affected by the Emerging Economic Crisis: The Case of Pregnant Indonesian Women." *European Journal of Clinical Nutrition* 57:654–66.

Hartsock, N. 1990. "Foucault on Power: A Theory for Women?" In *Feminism/Postmodernism,* ed. L. J. Nicholson, 157–75. New York: Routledge.

Hasan, Ibrahim. 1993. "Sambutan Pengarahan Menteri Negara Urusan Pangan/Ketua BULOG Widyakarya Nasional Pangan Dan Gizi V, Jakarta, 20 April 1993." In *Widyakarya Pangan Dan Gizi V,* ed. Mien A. Rifai, A. Nontji, Erwindodo, F. Jalal, D. Fardiaz, and T. Fallah, 15-20. Jakarta: Lembaga Ilmu Pengetahuan Indonesia.

Hays, Sharon. 1996. *The Cultural Contradictions of Motherhood.* New Haven: Yale University Press.

Headey, Derek, and Shenggen Fan. 2008. "Anatomy of a Crisis: The Causes and Consequences of Surging Food Prices." *Agricultural Economics* 39:375–91.

Heasman, Michael, and Julian Mellentin. 2001. *The Functional Foods Revolution: Healthy People, Healthy Profits?* London: Earthscan.

Heavy, Susan. 2008. "FDA Warns over Diet Coke Plus Nutrition Claims." Reuters, 23 December. http://www.reuters.com/article/2008/12/23/us-cocacola-fda-idUSTRE4BM49W20081223.

Hecht, Gabrielle. 2000. *The Radiance of France: Nuclear Power and National Identity after World War II.* Cambridge: MIT Press.

Helen Keller International. 2000. "Vitamin A Capsules: Red and Blue, What's the Difference?" *Indonesia Crisis Bulletin* 5 [accessed 10 May 2012]. http://www.hki.org/research/Ind%20Cris%20Bul%20y2%20iss%205.pdf.

Henderson, E. 2000. "Rebuilding Local Food Systems from the Grass-Roots Up." In *Hungry for Profit: The Agribusiness Threat to Farmers, Food, and the Environment,* ed. F. Magdoff, Bellamy J. Foster, and F. H. Buttel, 175–88. New York: Monthly Review Press.

Heringa, Rens. 1997. "Dewi Sri in Village Garb: Fertility, Myth, and Ritual in Northeast Java." *Asian Folklore Studies* 56:355–77.

Herman, Susilowati. 2002. "Comments on the Paper 'Increasing the Micronutrient Content of Rice through Plant Breeding.'" At the *Biofortification Seminar: Breeding for Micronutrient-Dense Rice to Complement Other Strategies for Reducing Malnutrition,* Ministry of Agriculture, 12 June, Bogor, Indonesia.

Hinrichs, C. Clare, and Thomas A. Lyson. 2008. *Remaking the North American Food System: Strategies for Sustainability.* Lincoln: University of Nebraska Press.

Holt-Gimenez, H., Miguel A. Altieri, and P. Rosset. 2006. "Ten Reasons Why the Rock-efeller and the Bill and Melinda Gates Foundations' Alliance for Another Green Revolution Will Not Solve the Problems of Poverty and Hunger in Sub-Saharan Africa." Food First/Institute for Food and Development Policy, Policy Brief, no. 12. October.

Hossain, Tahzeeba, Irwin Rosenberg, Jacob Selhub, Ganesh Kishore, Roger Beachy, and Karel Schubert. 2004. "Enhancement of Folates in Plants through Metabolic Engineering." *PNAS* 101, no. 14: 5158–63.

Howard, Mary, and Ann Millard. 1997. *Hunger and Shame: Child Malnutrition and Poverty on Mount Kilimanjaro.* London: Routledge.

Howe, Leo. 1991. "Rice, Ideology, and the Legitimation of Hierarchy in Bali." *Man* 26, no. 3: 445–67.

Hull, Terence H., and Valerie J. Hull. 2005. "From Family Planning to Reproductive Health Care: A Brief History." In *People, Population, and Policy in Indonesia,* ed. Terence H. Hull, 1–70. Jakarta: Equinox (Asia).

Hull, Valerie J. 1979. "Women, Doctors, and Family Health Care: Some Lessons from Rural Java." *Studies in Family Planning* 10, no. 11–12: 313–25.

Hunt, J. M. 2001a. "Asia Regional Actions, Strategic Priorities for Controlling Anemia" [accessed 3 March 2004]. http://www.idpas.org/pdf/2046AsiaRegionalActions.pdf.

———. 2001b. "Food Policy and Nutrition Security: Lessons Learned and New Paradigms." Paper presented at the Asia and Pacific Forum on Poverty: Reforming Policies and Institutions for Poverty Reduction, 5–9 February, Manila, Philippines.

———. 2002. "Kepentingan Investasi Pemerintah Dan Swasta Dalam Mengatasi Defisinsi Gizi Mikro." In *Fortifikasi Tepung Terigu Dan Minyak Goreng,* ed. Hardinsyah, Leily Amalia, and Budi Setiawan. Jakarta: Pusat Studi Kebijakan Pangan dan Gizi, IPB, Komisi Fortifikasi Nasional, ADB-Manila, and Keystone Center USA.

Hunt, J. M., and M. G. Quibria. 1999. "Preface." *Asian Development Review* 17, no. 1–2, i–iii.

Husaini, M. A., H. D. Karyadi, and H. Gunadi. 1981. "Evaluation of Nutritional Ane-mia Intervention among Anemic Female Workers on a Tea Plantation." In *Iron Deficiency and Work Performance,* ed. L. Hallberg and N. S. Scrimshaw, 72–85. Washington, DC: Nutrition Foundation.

Igun, U. A. 1982. "Child-Feeding Habits in a Situation of Social Change: The Case of Maiduguri, Nigeria." *Social Science and Medicine* 16:769–81.

IMF (International Monetary Fund). 1998. "Letter of Intent for IMF by the Govern-ment of Indonesia, dated September 11" [accessed 4 July 2005]. http://www.imf.org/external/np/loi/091198.HTM.

Indofood. 2003. *Annual Report 2003: Facing the Future.* Jakarta: Indofood.

INSTATE Pty Ltd. 2003. "*Food Exporters' Guide to Indonesia: A Report Prepared for the Australian Government Department of Agriculture, Fisheries and Forestry*" [accessed 12 December 2004]. www.daff.gov.au/foodinfo.

International Baby Food Action Network (IBFAN). 2004. *Breaking the Rules: Stretching the Rules.* Penang, Malaysia: IBFAN.

International Center for Tropical Agriculture and International Food Policy Research Insti-tute. 2002. "Biofortified Crops for Improved Human Nutrition: Challenge Program Proposal" [accessed 28 July 2004]. http://www.cgiar.org/pdf/biofortification.pdf.

International Zinc Nutrition Consultative Group. 2004. "Developing Zinc Interven-tion Programs." *Food and Nutrition Bulletin* 24, no. 1: 163–86.

IRIN News. 2011. "INDONESIA: Breastfeeding Regulations to Target Formula Com-panies" [accessed 1 June 2012]. http://www.irinnews.org/Report/93744/INDONESIA-Breastfeeding-regulations-to-target-formula-companies.

IRRI (International Rice Research Institute), Rockefeller Foundation, and Syngenta. 2001. "International Rice Research Institute Begins Testing 'Golden Rice.'" Press release, 22 January.

ISAAA (International Service for the Acquisition of Agri-biotech Applications). 2010. "Global Status of Commercialized Biotech/GM Crops: 2010." ISAAA Brief 42 [accessed 2 December 2011]. http://www.isaaa.org/resources/publications/briefs/42/executivesummary/default.asp.

——. 2011. "Global Status of Commercialized Biotech/GM Crops 2011" [accessed 16 May 2012]. http://www.isaaa.org/resources/publications/briefs/43/executivesummary/default.asp.

ISAAA, and University of Illinois at Urbana–Champaign. 2002. "The Social and Cultural Dimensions of Agricultural Biotechnology in Southeast Asia: Public Understanding, Perceptions, and Attitudes towards Biotechnology in Indonesia" [accessed 3 May 2004]. http://www.isaaa.org/kc/inforesources/publications/perception/Indonesia.pdf.

Jaffe, Daniel. 2007. *Brewing Justice: Fair Trade Coffee, Sustainability, and Survival.* Berkeley: University of California Press.

Jahan, Rounaq. 1995. *The Elusive Agenda: Mainstreaming Women in Development.* Atlantic Highlands, NJ: Zed Books.

Jakarta Post. 2000. "Bulog Scandal Probe Might Hurt Golkar." 27 June.

——. 2001a. "Formula Milk 'Not against Breastfeeding.'" 21 October.

——. 2001b. "Making Breast-Feeding a Right." 16 September.

——. 2001c. "Mothers Target of Battle over Bottle." 24 June.

——. 2001d. "President Reminded of Her Promise on GMO." 8 September.

——. 2002. "Bt Cotton in Indonesia Has Contaminated Natural Cotton." 21 November.

——. 2009. "Child Dies of Malnutrition in East Nusa Tenggara." 11 January.

Jalal, Fasli, and Sumali M. Atmojo. 1998. "Peranan Fortifikasi Dalam Penanggulangan Masalah Kekurangan Zat Gizi Mikro." In *Widyakarya Pangan Dan Gizi VI,* ed. F. G. Winarno, 915–38. Jakarta: Lembaga Ilmu Pengetahuan Indonesia.

Jasanoff, Sheila. 2003. "Technologies of Humility: Citizen Participation in Governing Science." *Minerva* 41:223–44.

Joint FAO/WHO Expert Committee on Nutrition. 1952. *Report on the Third Session.* Geneva: WHO.

Kabeer, Naila. 1999. "Resources, Agency, Achievements: Reflections on the Measurement of Women's Empowerment," *Development and Change* 30, no. 3: 435–64.

Karyadi, Darwin. 1973a. "Hubungan Ketahanan Fisik Dengan Keadaan Gizi Dan Anemi Gizi Besi," PhD diss., University of Indonesia.

——. 1973b. "Nutrition and Health of Indonesian Construction Workers." Staff Working Paper, International Bank for Reconstruction and Development, no. 152, prepared by Darwin Karyadi and Samir Basta. Washington, DC: International Bank for Reconstruction and Development.

Kay, Lily E. 2000. *Who Wrote the Book of Life?: A History of the Genetic Code.* Stanford: Stanford University Press.

Kennedy, Gina, Guy Nantel, and Prakash Shetty. 2003. "The Scourge of 'Hidden Hunger': Global Dimensions of Micronutrient Deficiencies." *Food, Nutrition and Agriculture* 32:8–16.

Kenney, R. A., and C. S. Tidball. 1972. "Human Susceptibility to Oral Monosodium L-Glutamate." *American Journal of Clinical Nutrition* 25:140–46.

Kesmodel, D., L. Etter, and, A. Patrick. 2008. "Grain Companies' Profits Soar as Global Food Crisis Mounts." *Wall Street Journal,* 30 April.

Kickbusch, Ilona. 2000. "The Development of International Health Policies— Accountability Intact?" *Social Science and Medicine* 51, no. 6: 979–89.

Kiess, Linnda, Regina Moench-Phanner, M. W. Bloem, Saskia dePee, Mayang Sari, and Soewarta Kosen. 2000. "New Conceptual Thinking about Surveillance: Using Micronutrient Status to Assess the Impact of Economic Crisis on Health and Nutrition." *Malaysian Journal of Nutrition* 6, no. 2: 223–32.

Kilaru, Asha, Zoe Matthews, Jayashree Ramakrishna, Shanti Mahendra, and Saraswathy Ganapathy. 2004. "'She has a tender body': Postpartum Morbidity and Care during Banathana in Rural South India." In *Reproductive Agency, Medicine, and the State: Cultural Transformations in Childbearing,* ed. M.Unnithan-Kumar, 161-80. New York: Berghahn Books.

Kimura, Aya Hirata. 2010. "Between Technocracy and Democracy: An Experimental Approach to Certification of Food Products by Japanese Consumer Cooperative Women." *Journal of Rural Studies* 26, no.2: 130–50.

——. 2011. "Food Education as Food Literacy: Privatized and Gendered Food Knowledge in Contemporary Japan" *Agriculture and Human Values* 28, no. 4: 465–82.

King, Janet C. 2002. "Evaluating the Impact of Plant Biofortification on Human Nutrition." *Journal of Nutrition* 132, no. 3: S511–13.

Kjaernes, Unni. 1995. "Political Struggle over Scientific Definitions: Nutrition as a Social Problem in Interwar Norwegian Nutrition Policy." In *Eating Agendas: Food and Nutrition as Social Problems,* ed. Donna Maurer and Jeffery Sobal, 261–78. New York: Aldine de Gruyter.

Kleinman, Daniel Lee. "Democratizations of Science and Technology." In *Science, Technology, and Democracy,* ed. D. L. Kleinman,139–66. New York: State University of New York Press.

Kloppenburg, Jack, Jr. 2004. *First the Seed: The Political Economy of Plant Biotechnology, 1492–2000.* Madison: University of Wisconsin Press.

Kloppenburg, Jack, Jr., and Sharon Lezberg. 1996. "Getting It Straight before We Eat Ourselves to Death: From Food System to Foodshed in the 21st Century." *Society and Natural Resources* 9: 93–96.

Komari. 2000. "Makanan Pendamping Asi (MP-ASI) Sebagai Teknologi Intervensi Gizi." *Kumpulan Makalah Diskusi Pakan Bidang Gizi Tentang ASI-MP AS.* Jakarta: Persatuan Ahli Gizi Indonesia, Lembaga Ilmu Pengetahuan Indonesia, UNICEF.

Komari, and Hermana. 1993. "Fortifikasi Zat Besi Pada Tepung Terigu Dan Kecap." *Penelitian Gizi Dan Makanan* 16:113–16.

Kompas. 2003a. "Aptindo Khawatir Terigu Impor Tak Ber-SNI Terus Membanjiri Pasar Lokal." 11 September.

——. 2003b. "Fortifikasi Terigu Impor Dinilai Sebagai Hambatan Dagang." 27 September.

——. 2003c. "Komoditas Tepung Terigu Impor Wajib Didaftarkan Di Depkes." 18 July.

Kosen, Seowarta, Benny Kodyat, Muhilal, D. Karyadi, and Rainer Gross. 1998. "Suggested Actions for Iron Deficiency Control in Indonesia." *Nutrition Research* 18, no. 12: 1965–71.

Kristof, Nicholas. 2009. "The Hidden Hunger." *New York Times,* 24 May.

Kurniawan, A. 2002. "Policies in Alleviating Micronutrient Deficiencies: Indonesia's Experience." *Asia Pacific Journal of Clinical Nutrition* 11, no. 3: S360–70.

Kwok, R. H. M. 1968. "Chinese-Restaurant Syndrome." *New England Journal of Medicine* 278: 96.

Kwok, Yenni. 1997. "How to Grow a Market: Removing Controls on Wheat Is a Big Step." *Asia Week* [accessed 20 July 2004]. http://www-cgi.cnn.com/ASIANOW/asiaweek/97/1114/biz2.html.

Ladd-Taylor, Molly. 2004. "Mother-Worship/Mother Blame: Politics and Welfare in an Uncertain Age." *Journal of the Association for Research on Mothering* 6, no. 1: 7–15.

Lancet. 1990. "Editorial: Structural Adjustment and Health in Africa." *Lancet* 335: 885–86.

Lang, Tim. 1999. "Diet, Health and Globalisation: Five Key Questions." *Proceedings of the Nutrition Society* 58:335–43.

———. 2010. "Crisis? What Crisis?: The Normality of the Current Food Crisis." *Journal of Agrarian Change* 10, no. 1: 87–97.

Lang, Tim, and Michael Heasman. 2004. *Food Wars.* London: Earthscan.

Lappé, Frances Moore, and Anna Lappé. 2002. *Hope's Edge: The Next Diet for a Small Planet.* New York: Tarcher.

Latour, Bruno. 1987. *Science in Action: How to Follow Scientists and Engineers through Society.* Milton Keynes, UK: Open University Press.

Lee, Kelley, Sue Collinson, Gill Walt, and Lucy Gilson. 1996. "Who Should Be Doing What in International Health: A Confusion of Mandates in the United Nations?" *British Medical Journal* 312, no. 7026: 302.

Lehmann, V. 2001. "Biotechnology in the Rockefeller Foundation's New Course of Action." *Biotechnology and Development Monitor* 44–45:15–19. http://www.biotech-onitor.nl/4406.htm.

Levenstein, Harvey A. 1993. *Paradox of Plenty: A Social History of Eating in Modern America.* New York: Oxford University Press.

———. 2003. *Revolution at the Table: The Transformation of the American Diet.* Berkeley: University of California Press.

Levine, Susan. 2008. *School Lunch Politics: The Surprising History of America's Favorite Welfare Program.* Princeton: Princeton University Press.

Levinson, F. James. 1993. *Incorporating Nutrition into Bank-Assisted Social Funds.* World Bank Human Resources Development and Operations Policy (HRO) Working Paper, no. 5. Washington, DC: World Bank.

———. 1999. "Searching for a Home: The Institutionalization Issue in International Nutrition." Background paper, World Bank–UNICEF Nutrition Assessment. Washington, DC: World Bank and UNICEF.

Levinson, F. James, and Milla McLachlan. 1999. "How Did We Get Here?: A History of International Nutrition." *Scaling Up, Scaling Down Overcoming Malnutrition in Developing Countries,* ed. Thomas Marchione, 41–48. Amsterdam: Gordon and Breach.

Levi-Strauss, Claude. (1969) 1983. *The Raw and the Cooked.* Chicago: University of Chicago Press.

Li, Tania Murray. 2007. *The Will to Improve: Governmentality, Development, and the Practice of Politics.* Durham, NC: Duke University Press.

Liddle, R. William. 1999. "Indonesia's Unexpected Failure of Leadership." In *The Politics of Post-Suharto Indonesia,* ed. Adam Schwarz and Jonathan Paris, 16–39. New York: Council on Foreign Relations Press.

Litt, Jacquelyn S. 2000. *Medicalized Motherhood: Perspectives from the Lives of African-American and Jewish Women.* New Brunswick, NJ: Rutgers University Press.

Lock, Margaret, and Patricia A. Kaufert. 1998. *Pragmatic Women and Body Politics.* Cambridge: Cambridge University Press.

Lynch, Sean. 2005. "The Impact of Iron Fortification on Nutritional Anemia." *Best Practice and Research Clinical Haematology* 18, no. 2: 333–46.

Lyson, Thomas A. 2004. *Civic Agriculture: Reconnecting Farm, Food, and Community.* Medford, MA: Tufts University Press.

Maberly, Glen F. 2002. "A Policy Planning Forum with the Wheat Industry to Explore the Launch of Global Public-Private Initiative Supporting Universal Wheat Fortification" [accessed 14 May 2012]. http://www.sph.emory.edu/wheatflour/Comm/Resource/CDs/Mauritius/files/Flour%20Fortification%20Paper%20October%209.pdf.

Maberly, Glenn F., F. L. Trowbridge, and K. M. Sullivan. 1994. "Programs against Micronutrient Malnutrition: Ending Hidden Hunger." *Annual Review of Public Health* 15: 277–301.

MacCormack, Carol. 1981. "Health Care and the Concept of Legitimacy." *Social Science and Medicine* 15, no. 3: 423–28.

——. 1986. "The Articulation of Western and Traditional Systems of Health Care." In *The Professionalization of African Medicine,* ed. G. L. Chavunduka and Last Murray. Manchester, UK: Manchester University Press.

Madden, Normandy. 2003. "Formula Marketer Busts Myths." *Advertising Age,* 13 October.

Madeleine, J. 2000. "Grains of Hope." *Time,* 31 July.

Magdoff, Fred, John Bellamy Foster, and Frederick H Buttel. 2000. *Hungry for Profit: The Agribusiness Threat to Farmers, Food, and the Environment.* New York: Monthly Review Press.

Magiera, Stephen L. 1993. "Grain Quality as a Determinant of Wheat Import Demand: Case of Indonesia." A Report for USAID. Jakarta: USAID.

Manila Times. 2011. "Lawmakers, Farmers, Consumers Question Safety of Golden Rice." 4 September.

Manning, Chris. 1988. "Rural Employment Creation in Java: Lessons from the Green Revolution and Oil Boom." *Population and Development Review* 14, no. 1: 47–80.

Margolis, Maxine L. 1984. *Mothers and Such: Views of American Women and Why They Changed.* Berkeley: University of California Press.

Martoatmodjo, Soekartijah, Djumadias Abu Nain, Muhilal, Muhammad Enoch, and Soemilah Husaini./Sastroamidjojo. 1973. "Masalah Anemi Gizi Pada Wanita Hamil Dalam Hubungannya Dengan Pola Konsumsi Makanan." *Penelitian Gizi Dan Makanan* 3:22–41.

Martoatmodjo, Soekartijah, Muhilal, Muhammad Enoch, Husaini, W. Angkuw, and Dradjat D. Prawiranegara. 1972. "Anemi Gizi Pada Wanita Hamil di Desa Bendungan, Kabupaten Bogor." *Penelitian Gizi Dan Makanan* 2:11–23.

Martoatmodjo, Soekartijah, Muhilal, Soemilah Sastroamidjojo, and Iman Soemarno. 1980. "Pencegahan Anemi Gizi Besi Pada Kehamilan Dengan Suplemen Pil Sulfas Ferrosus Melalui Pusat Kesehatan Masyarakat." *Penelitian Gizi Dan Makanan* 4: 3–13.

Maslow, A. 1943. "A Theory of Human Motivation." *Psychological Review* 50:370–96.

McAfee, Kathleen. 2003. "Neoliberalism on the Molecular Scale: Economic and Genetic Reductionism in Biotechnology Battles." *Geoforum* 34:203–19.

McLaren, Donald. 1963. "World Hunger: Some Misconceptions." *Lancet* 13:86–87.

——. 1966. "A Fresh Look at Protein-Calorie Malnutrition." *Lancet* 2:485–88.

McLester, James S. (1939) 1949. *Nutrition and Diet in Health and Disease.* Philadelphia: W. B. Saunders.

McMichael, Philip. 2005. "Global Development and the Corporate Food Regime." *Research in Rural Sociology and Development* 11:265–99.

——. 2009. "Food Regime Genealogy." *Journal of Peasant Studies* 36, no. 1: 139–69.

McNay, L. 1992. *Foucault and Feminism.* Cambridge: Polity Press.

Mebrahtu, Saba, David Pelletier, and Per Pinstrup-Andersen. 1995. "Agriculture and Nutrition." In *Child Growth and Nutrition in Developing Countries: Priorities for Action,* ed. Per Pinstrup-Andersen, David L Pelletier, and Harold Alderman, 220–42. Ithaca: Cornell University Press.

Mellanby, E., and H. N. Green. 1929. "Vitamin A as an Anti-Infective Agent: Its Use in the Treatment of Puerperal Septicemia." *British Medicine* 1:984–86.

Melse-Boonstra, A., S., E. Martini, S. Halati, M. Sari, S. Kosen, Muhilal, and M. W. Bolem. 2000. "The Potential of Various Foods to Serve as a Carrier for Micronutrient Fortification: Data from Remote Areas in Indonesia." *European Journal of Clinical Nutrition* 54:822–27.

Meyer, J., J. Boli, G. Thomas, and F. Ramirez. 1997. "World Society and the Nation-State." *American Journal of Sociology* 103, no. 1: 144–81.

Meylinah, Sugiarti. 2008. *Indonesia Grain and Feed Annual 2008 Grain Report.* Jakarta: USDA Foreign Agricultural Service.

Micronutrient Initiative (MI). 1997. Introduction to *Food Fortification to End Micronutrient Malnutrition: State of the Art,* 6-14. Report for the Satellite Conference of the 16th International Congress of Nutrition. International Congress of Nutrition, Food Fortification to End Micronutrient Malnutrition: State of the Art, 2 August, Hotel Intercontinental, Montreal.

Miller, F. 1977. "Knowledge and Power: Anthropology Policy Research and the Green Revolution." *American Ethnologist* 4: 190–98.

Miltz, M. 2011. "The Authoritarian Face of the 'Green Revolution': Rwanda Capitulates to Agribusiness" [accessed 3 February 2012]. http://www.grain.org/bulletin_board/entries/4322-the-authoritarian-face-of-the-green-revolution-rwanda-capitulates-to-agribusiness.

Ministry of Education and Culture (Indonesia). n.d. "Education Development in Indonesia" [accessed 16 May 2003]. http://www.ibe.unesco.org/International/Databanks/Dossiers/rindones.htm.

Ministry of Health (Indonesia). 1999. *Strategi Nasional PP-ASI.* Jakarta: PP-ASI

Ministry of Trade and Industry (Indonesia). 2003. "Pemerintah Naikkan BM Terigu Sementara Dari 0% Menjadi 5%." *Media Indag* (special issue on Indonesian products). Jakarta: Ministry of Trade and Industry.

Miraftab, Faranak. 2004. "Public-Private Partnerships: The Trojan Horse of Neoliberal Development?" *Journal of Planning Education and Research* 24, no. 1: 89–101.

Mitchell, Timothy. 2002. *Rule of Experts: Egypt, Techno-Politics, Modernity.* Berkeley: University of California Press.

Mohanty, C. 1991. "Under Western Eyes: Feminist Scholarship and Colonial Discourses." In *Third World Women and the Politics of Feminism,* ed. Ann Russo, Lourdes Torres, and Chandra Talpade Mohanty, 51–80. Bloomington: Indiana University Press.

Monastra, G., and L. Rossi. 2003. "Transgenic Foods as a Tool for Malnutrition Elimination and Their Impact on Agricultural Systems." *Rivista Biologia* 96, no. 3: 363–84.

Monsanto. 2001. *2000 Annual Report.* St. Louis: Monsanto Company.

Monsanto Biotechnology Knowledge Center. 2001. "Why Biotechnology Matters" [accessed 16 May 2006]. http://www.biotechknowledge.com/biotech/bbasics.nsf/why.html?OpenPage.

Moon, Suzanne M. 1998. "Takeoff or Self-Sufficiency?: Ideologies of Development in Indonesia, 1957–1961." *Technology and Culture* 39, no. 2: 187–212.

Muhilal, and Ance Murdiana. 1985. "Teknologi Fortifikasi MSG Dengan Vitamin A." *Penelitian Gizi Dan Makanan* 8:57–66.

Muhilal, Ance Murdiana, and Hermana. 1997. "Terigu Sebagai Wahana Fortifikasi Zat Besi." *Gizi Indonesia* 22, no. 1: 1–6.

Muhilal, D. Permesieh, Y. Idjaradimata, Muherdiyantingsih, and D. Karyadi. 1988. "Vitamin A-Fortified Monosodium Glutamate and Health, Growth and Survival of Children: A Controlled Trial." *American Journal of Clinical Nutrition* 48:1271–76.

Murray, C. J. L., and A. D. Lopez. 1999. *The Global Burden of Disease*. Cambridge: Harvard University Press.

Mutersbaugh, T. 2005. "Fighting Standards with Standards: Harmonization, Rents, and Social Accountability in Certified Agrofood Networks." *Environment and Planning A* 37:2033–51.

Nagle, James J. 1976. "US Industry Set to Feed the Poor." *New York Times*, 13 August.

Nain, Djumadias Abu, and F. J. Maspaitella. 1973. "Pola Pemberian Makanan Kepada Bayi Di Beberapa Daerah Indonesia." *Penelitian Gizi Dan Makanan* 3:42–48.

Natakusuma, Suroso. 1997. "Kebijaksanaan Fortifikasi Produk Pangan Dalam Rangka Peningkatan Sumber Daya Manusia." Paper presented at Sarasehan Sehari Tentang Fortifikasi Produk Pangan, 12 March, Jakarta.

———. 1998. "Strategi Fortifikasi Pangan." In *Widyakarya Pangan Dan Gizi VI*, ed. F. G. Winarno, 901–8. Jakarta: Lembaga Ilmu Pengetahuan Indonesia.

National Academy of Sciences. Subcommittee on Vitamin A Deficiency Prevention and Control. Committee on International Nutrition Programs. Food and Nutrition Board. Commission on Life Sciences. National Research Council.1987. *Vitamin A Supplementation: Methodologies for Field Trials*. Washington, DC: National Academy Press.

Nestle, Marion. 2001. "Genetically Engineered 'Golden' Rice Unlikely to Overcome Vitamin A Deficiency." *Journal of American Dietetic Association* 101, no. 3: 289–90.

———. 2002. *Food Politics: How the Food Industry Influences Nutrition and Health*. California Studies in Food and Culture. Berkeley: University of California Press.

Newland, Lynda. 2001. "The Deployment of the Prosperous Family: Family Planning in West Java." *NWSA Journal* 13, no. 3: 22–48.

Office of the Minister of State for the Role of Women (Indonesia). 1990. "Protecting, Promoting and Supporting Breast-Feeding: Indonesian Experience." Jakarta: Office of the Minister of State for the Role of Women.

Okie, Susan. 2006. "Global Health: The Gates-Buffet Effect." *New England Journal of Medicine* 355, no. 11: 1084–88.

Oliver, J. Eric. 2006. *Fat Politics: The Real Story behind America's Obesity Epidemic*. New York: Oxford University Press.

Omawale. 1984. "Incorporating Nutritional Concerns into the Specification of Desired Technology Characteristics in International Agricultural Research." In *International Agricultural Research and Human Nutrition*, ed. Per Pinstrup-Andersen, Alan Berg, and Martin Forman, 265–75. Washington, DC: IFPRI.

Ong, Aihwa. 2000. "Graduated Sovereignty in South-East Asia." *Theory, Culture, & Society* 15, no. 4: 55–75.

———. 2002. "Globalization and New Strategies of Ruling in Developing Countries" *Études Rurales* 163–164:233–48.

———. 2011."Translating Gender Justice in Southeast Asia: Situated Ethics, NGOs, and Bio-Welfare." *Journal of Women of the Middle East and the Islamic World* 9: 26–48.

Ong, Aihwa, and S. Collier. 2004. *Global Assemblages: Technology, Politics, and Ethics as Anthropological Problems*. Oxford: Wiley-Blackwell.

Oomen, H. P., D. S. McLaren, and H. Escapini. 1964. "Epidemiology and Public Health Aspects of Hypovitaminosis A: A Global Survey on Xerophthalmia." *Tropical and Geographical Medicine* 4:271–315.

Opportunities for Micronutrient Interventions (OMNI). n.d. "The Program Against Micronutrient Malnutrition (PAMM)" [accessed 16 May 2003]. http://www.jsi.com/intl/omni/pamm.htm.

O'Reilly, Andrea. 2004. *From Motherhood to Mothering: The Legacy of Adrienne Rich's Of Woman Born.* New York: State University of New York Press.

Orlove, B. S. 1997. "Meat and Strength: The Moral Economy of a Food Riot." *Cultural Anthropology* 12, no. 2: 234–68.

PT Sriboga Raturaya. n.d. "Wheat Flour Fortified with Vitamins: Built to Improve the Health of the Nation." Brochure. Semarang, Indonesia: PT Sriboga Raturaya.

Paarlberg, Robert L. 2003. "Reinvigorating Genetically Modified Crops." *Issues in Science and Technology* [accessed 3 August 2005]. http://www.issues.org/19.3/paarlberg.htm.

Pachico, Douglas H. 1984. "Nutritional Objectives in Agricultural Research—the Case of CIAT." In *International Agricultural Research and Human Nutrition,* ed. Per Pinstrup-Andersen, Alan Berg, Martin Forman, 25–40. Washington, DC: IFPRI.

Paltrow, Lynn M. 1999. "Punishment and Prejudice: Judging Drug-Using Pregnant Women." In *Mother Trouble,* ed. Julia E. Hanigsberg and Sara Ruddick, 59–78. Boston: Beacon.

Pandi, E., and P. Srihartati. 1987. "Peranan Keluarga Berencana Dalam Menunjang Program Gizi." In *Prosiding Kursus Penegar Ilmu Gizi Dan Kongres VII Persagi, 25–27 November 1986,* 121–29. Jakarta: Persatuan Ahli Gizi Indonesia.

Pardey, Philip G., Julian M. Alston, and Vincent H. Smith. 1997. "Financing Science for Global Food Security." *IFPRI Annual Report 1997.* Washington, DC: IFPRI.

Park, Youngmee K., C. T. Barton, C. N. Sempos, E.Vanderveen, and Elizabeth A.Yetley. 2000. "Effectiveness of Food Fortification in the United States: The Case of Pellagra." *American Journal of Public Health* 90, no.5: 727–38.

Parkin, Katherine J. 2006. *Food Is Love: Advertising and Gender Roles in Modern America.* Philadelphia: University of Pennsylvania Press.

Parpart, Jane L. 1995. "Deconstructing the Development 'Expert': Gender, Development and the 'Vulnerable Groups.'" In *Feminism/Postmodernism/Development,* ed. M. H. Marchand and J. L. Parpart, 221–43. London: Routledge.

Parpart, Jane L., Patricia Connelly, and Eudine Barriteau. 2000. *Theoretical Perspectives on Gender and Development.* Ottawa: International Development Research Centre.

Patel, Rajeev. 2007. "Transgressing Rights: La Via Campesina's Call for Food Sovereignty." *Feminist Economics* 13, no. 1: 87–116.

Paxson, Heather. 2004. *Making Modern Mothers: Ethics and Family Planning in Urban Greece.* Berkeley: University of California Press.

Peck, J., and A. Tickell. 2002. "Neoliberalizing Space." *Antipode* 34: 380–404.

Peluso, Nancy Lee, and Peter Vandergeest. 2001. "Genealogies of the Political Forest and Customary Rights in Indonesia, Malaysia, and Thailand." *Journal of Asian Studies* 60, no. 3: 761–812.

Perhac, R. M. 1996. "Defining Risk: Normative Considerations." *Human and Ecological Risk Assessment* 2, no. 2: 381–92.

Perkins, John H. 1997. *Geopolitics and the Green Revolution: Wheat, Genes, and the Cold War.* Oxford: Oxford University Press.

Permaesih, D., Ance M. Dahro, and Hadi Riyadi. 1988. "Hubungan Status Anemi Dan Status Besi Wanita Remaja Santri." *Penelitian Gizi Dan Makanan* 11:38–46.

Petryna, Adriana. 2002. *Life Exposed: Biological Citizens after Chernobyl.* Princeton: Princeton University Press.

Pilcher, Jeffrey M. 1998. *¡Que Vivan los Tamales!: Food and the Making of Mexican Identity.* Albuquerque: University of New Mexico Press.

Pinstrup-Andersen, Per. 2000. "Improving Human Nutrition through Agricultural Research: Overview and Objectives." *Food and Nutrition Bulletin* 21, no. 4: 352–55.

Poletti, Susanna, Wilhelm Gruissem, and Christof Sautter. 2004. "The Nutritional Forti-
     fication of Cereals." *Current Opinion in Biotechnology* 15, no. 2: 162–65.
Pollan, Michael. 2008. *In Defense of Food.* New York: Penguin.
Pollard, Richard, and Michael Favin. 1997. "Social Marketing of Vitamin A in Three
     Asian Countries" [accessed 14 May 2012]. http://www.manoffgroup.com/
     resources/SocMarketA.pdf.
Potrykus, Ingo. 2004. "Experience from the Humanitarian Golden Rice Project: Extreme
     Precautionary Regulation Prevents Use of Green Biotechnology in Public
     Projects" [accessed 19 July 2005]. http://www.agbioworld.org/biotech-info/
     articles/biotech-art/potrykus.html.
Powell, Maria, and D. L. Kleinman. 2008. "Building Citizen Capacities for Participa-
     tion in Technoscientific Decision Making: The Democratic Virtues of the Con-
     sensus Conference Model." *Public Understanding of Science* 17:329–48.
Pratt, G. 2004. *Working Feminism.* Philadelphia: Temple University Press.
Pretorius, P. J., and Z. M. Smith. 1968. "The Effects of Various Skimmed Milk Formu-
     lae on the Diarrhea, Nitrogen Retention and Initiation of Cure in Kwashiorkor."
     Journal of Tropical Pediatrics 4: 50–60.
Pritchard, B. 2009. "The Long Hangover from the Second Food Regime: A World-
     Historical Interpretation of the Collapse of the WTO Doha Round." *Agriculture
     and Human Values* 26, no. 4: 297–307.
Program for Appropriate Technology for Health (PATH). 2000. "Ultra Rice
     Technology." Seattle: PATH.
Purnama, Philip S. 2000. "Flour Fortification with Iron in Indonesia." *Manila Forum
     2000: Strategies to Fortify Essential Foods in Asia and the Pacific 71.* Nutrition
     and Development Series. Manila: ADB.
——. 2002. "Pengalaman Fortifikasi Tepung Terigu Di Indonesia." In *Fortifikasi
     Tepung Terigu Dan Minyak Goreng,* ed. Hardinsyah, Leily Amalia, and Budi
     Setiawan, 49-53. Jakarta: Pusat Studi Kebijakan Pangan dan Gizi, IPB, Komisi
     Fortifikasi Nasional, ADB-Manila, and Keystone Center USA.
——. 2003. "Wheat Flour Fortification in Indonesia: Building Coalition to Fight
     Hidden-Hunger." Paper presented at the International Grains Council Confer-
     ence, 25 June, London.
Rabinow, Paul. 2003. *Anthropos Today.* Princeton: Princeton University Press.
Ramazanoglu, C., ed. 1993. *"Up against Foucault": Exploration of Some Tensions
     between Foucault and Feminism.* London: Routledge.
Rapp, Rayna. 1999. *Testing Women, Testing the Fetus: The Social Impact of Amniocentesis
     in America.* New York: Routledge.
Ratcliff, Katherine Strother. 2002. *Women and Health: Power, Technology, Inequality,
     and Conflict in a Gendered World.* Boston: Allyn and Bacon.
Raynolds, Laura. 1998. "Harnessing Women's Work: Restructuring Agricultural and Indus-
     trial Labor Forces in the Dominican Republic." *Economic Geography* 74, no. 2: 149.
——. 2002. "Wages for Wives: Renegotiating Gender and Production Relations in
     Contract Farming in the Dominican Republic." *World Development* 30, no. 5:
     783–98.
Reddy, Vinodini. 2002. "History of the International Vitamin A Consultative Group,
     1975–2000." *Journal of Nutrition* 132, no. 9: S2852-56.
Reinharz, Shulamit, and Lynn Davidman. 1992. *Feminist Methods in Social Research.*
     New York: Oxford University Press.
Reutlinger, Shlomo, and Marcelo Selowsky. 1975. "Undernutrition and Poverty."
     Washington, DC: International Bank for Reconstruction and Development.
Rice, Andrew. 2010. "The Peanut Solution." *New York Times,* 2 September.

Rice Today. 2003. "Steel Golden Rice?: March of Progress for Enhanced Nutrition." *Rice Today* 2, no. 2: 9.

Rich, Bruce. 1994. *Mortgaging the Earth: The World Bank, Environmental Impoverishment, and the Crisis of Development.* Boston: Beacon.

Rieff, David. 2008. "A Green Revolution for Africa?" *New York Times,* 12 October.

Rocha, Cecilia. 2001. "Urban Food Security Policy: The Case of Below Horizonte, Brazil." *Journal for the Study of Food and Society* 5, no. 1: 36–47.

Rogers, Beatrice. 2003. "Health and Economic Consequences of Malnutrition." In *Combating Malnutrition: Time to Act,* ed. Stuart Gillespie, Milla McLachlan, and Roger Shrimpton, 74–85. Washington, DC: World Bank.

Rohde, Jon. 1993. "Indonesia's Posyandus: Accomplishments and Future Challenges." In *Reaching Health for All,* ed. Jon Rohde, Meera Chatterjee, and David Morley, 135–57. New York: Oxford University Press.

Rose, Nikolas. 2007. *The Politics of Life Itself.* Princeton: Princeton University Press.

Rosegrant, M., and Hazell, P. 1999. *Transforming the Rural Asian Economy: The Unfinished Revolution.* Hong Kong: Oxford University Press for the ADBB.

Rosset, Peter M. 2006. *Food Is Different: Why We Must Get the WTO out of Agriculture.* London: Zed Books.

Rothman, Barbara Katz. 1989. *Recreating Motherhood Ideology and Technology in a Patriarchal Society.* New York: Norton.

Rovner, Sandy. 1986. "Saving the Lives of 4 Million Children." *Washington Post,* 23 December.

Rowe, G., and L. J. Frewer. 2005. "A Typology of Public Engagement Mechanisms." *Science, Technology, and Human Values* 30, no. 2: 251–90.

Ruger, J. P. 2005. "The Changing Role of the World Bank in Global Health." *American Journal of Public Health* 95, no 1: 60–70.

Ruxin, Joshua Nalibow. 1996. "Hunger, Science, and Politics: FAO, WHO, and UNICEF—Nutrition Policies, 1945–1978." London: University College.

Ryan, James G. 1984. "The Effects of the International Agricultural Research Centers on Human Nutrition—Catalo and Commentary." In *International Agricultural Research and Human Nutrition,* ed. Per Pinstrup-Andersen, Alan Berg, and Martin Forman, 199–223. Washington, DC: IFPRI.

Sachs, Carolyn. 1983. *Invisible Farmers: Women in Agricultural Production.* Totowa, NJ: Rowman and Allanheld.

———. 1996. *Gendered Fields: Rural Women, Agriculture, and Environment.* Boulder, CO: Westview Press.

Sahlins, Marshall. 1976. *Culture and Practical Reason.* Chicago: University of Chicago Press.

Saidin, M., Yusuf Mahmud, Moecherdiyantiningsih, Sukati, and Komala. 1995. "Efektifitas Fortifikasi Mie Instan Dengan Zat Besi Dan Vitamin A Terhadap Peningkata Kadar HB Dan Feretin Serum Ibu Hamil." *Penelitian Gizi Dan Makanan* 18:17–27.

Sarkar, Sahotra. 1998. *Genetics and Reductionism.* Cambridge: Cambridge University Press.

Sawicki, Jana. 1991. *Disciplining Foucault: Feminism, Power, and the Body.* New York: Routledge.

Schaumburg, H. H. Y., R. Byck, R. Gerstel, and J. H. Mashman. 1968. "Monosodium L-Glutamate: Its Pharmacology and Role in the Chinese Restaurant Syndrome." *Science* 163, no. 8: 26–28.

Schnapp, Nina, and Quirin Schiermeier. 2001. "Critics Claim 'Sight-Saving' Rice Is Over-Rated." *Nature* 410, no. 6828: 503.

Schuftan, Claudio. 1999. "A Different Challenge in Combating Micronutrient Deficiencies and Combating Protein Energy Malnutrition, or the Gap between

Nutrition Engineers and Nutrition Activists" [accessed 2 November 2005]. http://humaninfo.org/aviva/ch45.htm.

Schuftan, Claudio, V. Ramalingaswami, and F. Levinson. 1998. "Micronutrient Deficiencies and Protein-Energy Malnutrition." *Lancet* 351:1812.

Schultink, W., R. Gross, S. Sastroamidjojo, and D. Karyadi. 1996. "Micronutrients and Urban Life Style: Selected Studies in Jakarta." *Asia Pacific Journal of Clinical Nutrition* 5, no. 3: 145–48.

Schurman, Rachel, and Dennis D. Kelso. 2003. *Engineering Trouble: Biotechnology and Its Discontents.* Berkeley: University of California Press.

Schwarz, Adam, and Jonathan Friedland. 1991. "Indonesia: Empire of the Son." *Far Eastern Economic Review* 14 (March): 46-53

Sclove, Richard E. 2000. "Town Meetings on Technology: Consensus Conference as Democratic Participation." In *Science, Technology and Democracy,* ed. Daniel Lee Kleinman, 33–48. New York: State University of New York Press.

Scoones, Ian. 2002. *Agricultural Biotechnology and Food Security: Exploring the Debate.* IDS Working Paper, no. 145. Sussex, UK: Institute of Development Studies.

Scott, James C. 1998. *Seeing Like a State: How Certain Schemes to Improve the Human Condition Have Failed.* Yale Agrarian Studies, Yale ISPS Series. New Haven: Yale University Press.

Scott, Joan W. 1996. *Only Paradoxes to Offer: French Feminists and the Rights of Man.* Cambridge: Harvard University Press.

Scrinis, Gyorgy. 2008. "On the Ideology of Nutritionism." *Gastronomica* 8, no. 1: 39–48.

Semba, Richard D. 1999. "Vitamin A as 'Anti-Infective' Therapy, 1920–1940." *Journal of Nutrition* 129 (1999): 783–91.

———. 2001. "Nutrition and Development: A Historical Perspective." In *Nutrition and Health in Developing Countries,* ed. R. D. Semba and M. W. Bloem, 1–30. Totowa, NJ: Humana Press, 2001.

Sen, Amartya Kumar. 1981. *Poverty and Famines: An Essay on Entitlement and Deprivation.* Oxford: Clarendon Press.

Shapiro, Laura. 2009. *Perfection Salad: Women and Cooking at the Turn of the Century.* Berkeley: University of California Press.

Shaw, W. D., and C. P. Green. 1996. "Vitamin A Promotion in Indonesia: Scaling Up and Targeting Special Needs." *Strategies for Promoting Vitamin A Production, Consumption, and Supplementation: Four Case Studies,* ed. R.E. Seidel, 55–64. Washington, DC: Academy for Educational Development.

Shiva, Vandana. 1988. *Staying Alive: Women, Ecology and Survival in India.* New York: Zed Books.

———. 1991. *Violence of the Green Revolution: Third World Agriculture, Ecology, and Politics.* London: Zed Books.

———. 1997. *Biopiracy: The Plunder of Nature and Knowledge.* Cambridge: South End Press.

Shiva, Vandana, and Maria Mies. 1993. *Ecofeminism.* London: Zed Books.

Shivaramakrishnan, K. 2003. "Scientific Forestry and Genealogies of Development in Bengal." In *Nature in the Global South,* ed. Paul Greenough and A. L. Tsing, 253–85. Durham, NC: Duke University Press.

Sikkink, Kathryn. 1986. "Codes of Conduct for Transnational Corporations: The Case of the WHO/UNICEF Code." *International Organization* 40:815–40.

Silvey, Rachel. 2003. "Spaces of Protest: Gendered Migration, Social Networks, and Labor Activism in West Java, Indonesia." *Political Geography* 22:129–55.

Sjögren, Ebba, and Claes-Fredrik Helgesson. 2007. "The Q(u)ALYfying Hand: Health Economics and Medicine in the Shaping of Swedish Markets for Subsidized Pharmaceuticals." *Sociological Review* 55, no. 2: 215–40.

Smith, David Norman. 2007. "Faith, Reason, and Charisma: Rudolf Sohm, Max Weber, and the Theology of Grace." *Sociological Inquiry* 68, no. 3: 32–60.

Soekirman. 1974. *Priorities in Dealing with Nutrition Problems in Indonesia.* Ithaca: Cornell University Division of Nutritional Sciences.

———. 1998. "Tepatkah Dalam Krisis Moneter Ini Bicara Soal Fortifikasi Besi: Tidakkah Itu Masalah Kecil?" In *Widyakarya Pangan Dan Gizi VI,* ed. F. G. Winarno, 909–14. Jakarta: Lembaga Ilmu Pengetahuan Indonesia.

Soekirman, Atmarita, Abas B. Jahari, Sanjaya, and Drajat Martianto. 2005. "Review of Nutrition Situation, Conceptual Framework and Strategy for Nutrition Interventions in Indonesia: With Emphasis on Micronutrient Deficiencies." Draft of article, as of January 2005, available at Koalisi Fortifikasi Indonesia, Jakarta.

Soekirman, Satoto, Drajat Martianto, Abas B. Jahari, Atmarita, Venkatesh Mannar, and Geoffrey Marks. 2003. *Situational Analysis of Nutrition Problems in Indonesia: Its Policy, Programs and Prospective Development.* Jakarta: Ministry of Health, Directorate General of Community Health, Directorate of Community Nutrition, and the World Bank.

Soetrisno, Uken S., D. S. Slamet, and Hermana. 1991. "Fortifikasi Mi Dengan Zat Besi." *Gizi Indonesia* 14:116–20.

Solomons, Noel W. 1999. "Child Nutrition in Developing Countries and Its Policy Consequences." In *Food in Global History,* ed. Raymond Grew, 149–68. Boulder, CO: Westview Press.

Somantri, Ida Hanarida, and Siti Dewi Indrasari. 2002. "Breeding for Iron Dense Rice: A Low Cost, Sustainable Approach to Reducing Anemia in Asia." Jakarta: Indonesian Center for Agricultural Biotechnology and Genetic Resources Research and Development, Indonesian Institute for Rice Research, IFPRI, and Indonesian Center for Food Crops Research and Development.

———. 2003. "Breeding for Iron Dense Rice: A Low Cost, Sustainable Approach to Reducing Anemia in Asia." Indonesian Center for Agricultural Biotechnology and Genetic Resources Research and Development and Indonesian Institute for Rice Research, IFPRI, and Indonesian Center for Food Crops Research and Development.

———. 2004. "Breeding for Iron Dense Rice: A Low Cost, Sustainable Approach to Reducing Anemia in Asia." Indonesian Center for Agricultural Biotechnology and Genetic Resources Research and Development and Indonesian Institute for Rice Research, IFPRI, and Indonesian Center for Food Crops Research and Development.

Sommer, A., I. Tarwotjo, E. West, K. P. Djunaedi, A. A. Loeden, R. Tilden, L. Mele, and Aceh Study Group. 1986. "Impact of Vitamin A Supplementation on Childhood Mortality: A Randomized Controlled Community Trial." *Lancet* 1:169–73.

Spar, Debora. 1996. "Trade, Investment, and Labor: The Case of Indonesia." *Columbia Journal of World Business* 31, no. 4: 30–39.

Spivak, Gayatri C. 1999. *A Critique of Postcolonial Reason: Toward a History of the Vanishing Present.* Cambridge: Harvard University Press.

Sprague, Joey. 2005. *Feminist Methodologies for Critical Researchers.* Walnut Creek, CA: Rowman & Littlefield.

Stage, Sarah. 1997. "Home Economics, What's in the Name?" In *Rethinking Home Economics: Women and the History of a Profession,* ed. Sarah Stage and Virginia B. Vincenti, 1–14. Ithaca: Cornell University Press.

Star, Susan Leigh, and James R. Griesemer. 1999. "Institutional Ecology, 'Translations' and Boundary Objects: Amaterus and Professionals in Berkeley's Museum of Vertebrate Zoology, 1907–1939." In *The Science Studies Reader,* ed. Mario Biagioli, 305–524. New York: Routledge.

Stoler, Ann Laura. 1995. *Race and the Education of Desire: Foucault's History of Sexuality and the Colonial Order of Things*. Durham, NC: Duke University Press.

———. 2002. *Carnal Knowledge and Imperial Power: Race and the Intimate in Colonial Rule*. Berkeley: University of California Press.

Stone, Glenn Davis. 2002. "Both Sides New: Fallacies in the Genetic-Modification Wars, Implications for Developing Countries, and Anthropological Perspectives." *Current Anthropology* 43, no. 5: 611–30.

Suara Pembaruan Online. 1996. "Wapres Canangkan Gerakan Pekerja Wanita Sehat Dan Produktif." 14 November [accessed 6 December 2005]. http://www.suarapembaruan.com/News/1996/11/141196/Headline/04/04.html.

Sudjasmin, Sri Muljati, Sihadi, Suhartato, and M. A. Husaini. 1993. "Pola Menyusui Dan Pemberian Makanan Pada Anak Balita Penderita Gizi Buruk Di Wilayah Bogor." *Penelitian Gizi Dan Makanan* 16:22–28.

Sukati, S., Moecherdiyantiningsih, Sri Murni Prastowo, and Komala. 1995. "Dampak Fortifikasi Mie Instan Dengan Vitamin A Dan Zat Besi Terhadap Status Vitamin A Dan Status Besi Anak Balita." *Penelitian Gizi Dan Makanan* 18:38–39.

Sunawang, Yusrizal, Tjitjik Jusiani, and Ernest Schofellen. 2000. "Public Awareness and Participation in the IDD Elimination Program in Indonesia." *Gizi Indonesia:* 9–16.

Surbakti, Sudarti. 1987. "Integrasi Gizi Dalam Susenas." In *Prosiding Kursus Penegar Ilmu Gizi Dan Kongres VII Persagi, 25–27 November 1986, Jakarta*, 220–28. Jakarta: Persatuan Ahli Gizi Indonesia.

Tarnoff, Curt, and Larry Nowels. 2004. "Foreign Aid: An Introductory Overview of U.S. Programs and Policy: CRS Report to Congress." Washington, DC: Congressional Research Service.

Taylor, J., L. Layne, and D. Wozniak, eds. 2004. *Consuming Motherhood*. New Brunswick, NJ: Rutgers University Press.

TEMPO. 2005a. "Busung Lapar Bukan Mengada-Ada." *TEMPO*, 20 June.

———. 2005b. "Busung Lapar Di Lumbung Beras." *TEMPO*, 6 June.

Third World Network Malaysia. 2005. "Monsanto Bribery Charges in Indonesia by DoJ and USSEC" [accessed 30 January 2005]. http://www.mindfully.org/GE/2005/Monsanto-Indonesia-Bribery27jan05.htm.

Thoenes, P. 2004. *The Role of Soybean in Fighting World Hunger*. FAO Commodities and Trade Division, Basic Foodstuffs Service. Rome: FAO. http://www.fao.org/es/esc/common/ecg/125/en/The_role_of_soybeans.pdf.

Thompson, S. J., and J. T. Cowan. 2000. "Globalizing Agro-Food Systems in Asia: Introduction." *World Development* 28, no. 3: 401–7.

Thorbecke, Erik, and Theodore van der Pluijm. 1993. *Rural Indonesia: Socio-Economic Development in a Changing Environment*. New York: NYU Press for the International Fund for Agricultural Development.

Time. 2000. "This Rice Could Save a Million Kids a Year." *Time* 156, no. 5.

Timmer, C. Peter. 2004. *Food Security in Indonesia: Current Challenges and the Long-Run Outlook*. Working Paper, no. 48. Washington, DC: Center for Global Development.

Tiwon, Sylvia. 1999. "From Heroes to Rebels: Jakarta's Aceh Policy Suddenly Looks Remarkably Colonial." *Inside Indonesia* [accessed 14 May 2012]. http://www.insideindonesia.org/edition-62-apr-jun-2000/from-heroes-to-rebels-3007554.

Tjandraningsih, Indrasari. 2000. "Gendered Work and Labour Control: Women Factory Workers in Indonesia." *Asian Studies Review* 24, no. 2: 257–168.

Toenniessen, Gary, Akinwumi Adesina, and Joseph DeVries. 2008. "Building an Alliance for a Green Revolution in Africa." *Annals of the New York Academy of Sciences* 1136, no. 1: 233–42.

Tripp, Robert. 1984a. "Nutrition in Agricultural Research at CIMMYT." In *International Agricultural Research and Human Nutrition,* ed. Per Pinstrup-Andersen, Alan Berg, and Martin Forman, 41–55. Washington, DC: IFPRI.

——. 1984b. "Production Research at the International Agricultural Research Centers and Nutritional Goals." In *International Agricultural Research and Human Nutrition,* ed. Per Pinstrup-Andersen, Alan Berg, and Martin Forman, 283–93. Washington, DC: IFPRI.

——. 1990. "Does Nutrition Have a Place in Agricultural Research?" *Food Policy* 15, no. 6: 467–74.

True, J., and M. Mintrom. 2001. "Transnational Networks and Policy Diffusion: The Case of Gender Mainstreaming." *International Studies Quarterly* 45:27–57.

Tsing, Anna. 2004. *Friction: An Ethnography of Global Connection.* Princeton: Princeton University Press.

Turner, B. 1982. "The Government of the Body: Medical Regimens and the Rationalization of Diet." *British Journal of Sociology* 33, no. 2: 254–69.

Underwood, Barbara A. 1998. "Prevention of Vitamin A Deficiency." In *Prevention of Micronutrient Deficiencies: Tools for Policymakers and Public Health Workers,* ed. Christopher Paul Howson, Eileen T. Kennedy, A. Horwitz, 103–65. Washington, DC: National Academy Press.

——. 2004. "Vitamin A Deficiency Disorders: International Efforts to Control a Preventable 'Pox.'" *Journal of Nutrition* S134: S231–36.

Underwood, Barbara A., and Suttilak Smitasiri. 1990. "Micronutrient Malnutrition: Policies and Programs for Control and Their Implications." *Annual Review of Nutrition* 19:303–24.

UNICEF (United Nations Childrens Fund). 1989. *The State of the World's Children.* New York: Oxford University Press.

——. 2003. "Wheat Flour Fortification in Indonesia." Jakarta: UNICEF.

UNICEF and Micronutrient Initiative (MI). 2003. *Vitamin and Mineral Deficiency: A Global Progress Report.* Ottawa: UNICEF and MI.

United Nations, Administrative Committee on Coordination, Subcommittee on Nutrition (UN ACC/SCN). 1978. "Food and Nutrition Bulletin." *Food and Nutrition Bulletin* 1, no. 1.

——. 1997. *Third Report on the World Nutrition Situation: A Report Compiled from Information Available to the ACC/SCN.* Geneva: FAO

——. 1998. *Full Report of the Meeting of the Working Group on Iron Deficiency. Report of the Sub-Committee on Nutrition at Its Twenty-Fifth Session,* 30 March–2 April 1998. Administrative Committee on Coordination, (UN) B04, Oslo.

——. 2000. *Fourth Report on the World Nutrition Situation: Nutrition throughout the Life Cycle.* Geneva: ACC/SCN.

United Nations, Advisory Committee on the Application of Science and Technology to Development. 1968. *Action to Avert the Impending Protein Crisis.* New York: United Nations.

United Nations, Panel of Experts on the Protein Problem Confronting Developing Countries. 1971. *Strategy Statement on Action to Avert the Protein Crisis in the Developing Countries.* New York: United Nations.

United Nations. 1992a. *Basic Facts about the United Nations.* New York: United Nations

——. 1992b. *Second Report on the World Nutrition Situation.* Geneva: FAO.

Unnithan-Kumar, Maya. 2004. "Introduction: Reproductive Agency, Medicine and the State." In *Reproductive Agency Medicine, and the State: Cultural Transformations in Childbearing,* ed. Maya Unnithan-Kumar, 1–23. Oxford: Berghahn Books.

Untoro, Rachmi. 2002. "Masalah Gizi Mikro Di Indonesia Dan Potensi Penang-gulangannya." In *Fortifikasi Tepung Terigu Dan Minyak Goreng*, ed. Hardin-syah, Leily Amalia, and Budi Setiawan, 5-20. Jakarta: Pusat Studi Kebijakan Pangan dan Gizi, IPB, Komisi Fortifikasi Nasional, ADB-Manila, and Keystone Center USA.

USDA. 1997. "Indonesia Begins Liberalization Process in Wheat Sector" [accessed May 14, 2012]. http://www.fas.usda.gov/grain/circular/1997/97-12/grain_1.htm.

——. n.d. Production, Supply and Distribution Online data [accessed 6 December 2011]. http://www.fas.usda.gov/psdonline/.

Utomo, Budi. 2000. "The Slowing Progress of Breastfeeding Promotion in Indonesia: Causes and Recommendation." In *Kumpulan Makalah Diskusi Pakan Bidang Gizi Tentang ASI-MP ASI*. Jakarta: Persatuan Ahli Gizi Indonesia, Lembaga Ilmu Pengetahuan Indonesia, UNICEF.

van der Ploeg, Jan Douwe. 2010. "The Food Crisis, Industrialized Farming and the Imperial Regime." *Journal of Agrarian Change* 10, no. 1: 98–106.

Viteri, F. E. 1999. "Iron Supplementation as a Strategy for the Control of Iron Deficiency and Ferropenic Anemia." *Latin American Archives* of *Nutrition* 49, no. 2: S15–22.

Volti, Rudi. 1995. *Society and Technological Change*. New York: St. Martin's Press.

Walker, Ronald, and John R. Lupien. 2000. "The Safety Evaluation of Monosodium Glutamate." *Journal of Nutrition* 130, no. 4: 1049.

Waterlow, J. C., and P. R. Payne. 1975. "The Protein Gap." *Nature* 258:113–17.

Weber, Max. 1978. *Economy and Society*. Berkeley: University of California Press.

Weinberg, A. 1991. "Can Technology Replace Social Engineering?" In *Controlling Technology: Contemporary Issues*, ed. W. B. Thompson, 41–48. Buffalo, NY: Pro-metheus Books.

Welch, R. M. 2002. "Breeding Strategies for Biofortified Staple Plant Foods to Reduce Micronutrient Malnutrition Globally." *Journal of Nutrition* 132, no. 3: S495–99.

Wessing, Robert. 1990. "Sri and Sedana and Sita and Rama: Myths of Fertility and Generation." *Asian Folklore Studies* 49:235–57.

West, Keith P., Jr. 2002. "Extent of Vitamin A Deficiency among Preschool Children and Women of Reproductive Age." *Journal of Nutrition* 132, no. 9: S2857–66.

Whatmore, Sarah. 1990. *Farming Women: Gender, Work and Family Enterprise*. London: Macmillan.

White, Mary. 1990. "Improving the Welfare of Women Factory Workers: Lessons from Indonesia." *International Labor Review* 129, no. 1: 121–33.

Wie, T. K. 2002. "The Soeharto Era and After: Stability, Development and Crisis, 1966–2000." In *The Emergence of a National Economy in Indonesia, 1800–2000*, ed. H. W. Dick, V. J. H. Houben, J. Th. Lindblad, and Thee Kian Wie, 194–243. Sydney: Allen & Unwin.

Wield, David, Joanna Chataway, and Maurice Bolo. 2010. "Issues in the Political Economy of Agricultural Biotechnology." *Journal of Agrarian Change* 10, no. 3: 342–66.

Williams, C. D. 1935. "Kwashiorkor, a Nutritional Disease of Children Associated with a Maize Diet." *Lancet* 2:1151–52.

Winarto, Yunita. 2005. "Food Security and Food Sovereignty: Fight for Right in Discourse in Practice." Paper presented at the Symposium of the Journal *Antropologi Indonesia*, 12–15 July, Jakarta.

Windfuhr, Michael, and Jennie Jonsén. 2005. *Food Sovereignty: Towards Democracy in Localized Food Systems*. Warwickshire, UK: ITDG Publishing.

Woodhouse, Stephen J. 1999. Letter to PT Sriboga Raturaya, 7 December.

Worboys, Michael. 1988. "The Discovery of Colonial Malnutrition between the Wars." In *Imperial Medicine and Indigenous Societies,* ed. David Arnold, 208–25. Manchester, UK: Manchester University Press.

World Bank. n.d. "To Nourish a Nation: Investing in Nutrition with World Bank Assistance" [accessed 20 November 2003]. http://www.worldbank.org/html/extdr/hnp/nutritiontnan/htm.

———. 1994. *Enriching Lives: Overcoming Vitamin and Mineral Malnutrition in Developing Countries.* Washington, DC: World Bank.

———. 2007. *World Development Report 2008: Agriculture for Development.* Washington, DC: World Bank.

———. 2008a. *World Development Report 2007.* Washington, DC: World Bank.

———. 2008b. "World Bank President Calls for Plan to Fight Hunger in Pre-Spring Meetings Address." Press release, 2 April.

———. 2011. "World Bank HNP Lending" [accessed 11 December 2011]. www.worldbank.org.

World Health Organization (WHO). 1998. *Complementary Feeding of Young Children in Developing Countries: A Review of Current Scientific Knowledge.* Geneva: WHO.

———. 2001a. *Complementary Feeding: A Report of the Global Consultation.* Geneva: WHO.

———. 2001b. *Iron Deficiency Anaemia Assessment, Prevention and Control: A Guide for Programme Managers.* Geneva: WHO.

———. 2003. "Vitamin A supplementation" [accessed 16 May 2003]. http://www.who.int/vaccines/en/vitamina.shtml.

Wynne, Brian. 2001. "Creating Public Alienation: Expert Cultures of Risk and Ethics on GMOs." *Science as Culture* 10, no. 4: 445–81.

Ye, X., S. Al-Babili, A. Kloti, J. Zhang, P. Beyer, and I. Potrykus. 2000. "Engineering Provitamin A (b-Carotene) Biosynthetic Pathway into (Cartenoid Free) Rice Endosperm." *Science* 287:303–5.

Young, Iris M. 2000. *Inclusion and Democracy.* Oxford: Oxford University Press.

Zimmermann, Roukayatou, and Matin Qaim. "2004. Potential Health Benefits of Golden Rice: A Philippine Case Study." *Food Policy* 29, no. 2: 147–68.

Index

abstract individuation, 60

ACC/SCN (UN Administrative Committee on Coordination, Sub-Committee on Nutrition), 28, 49, 57

Aceh study, 28, 29–30, 69, 180n11

ADB. *See* Asian Development Bank (ADB)

Aditjondro, George, 86, 89

advertising. *See* marketing

Africa: AGRA (Alliance for a Green Revolution in Africa), 14, 15–16, 177n39, 188n9; agriculture in, 178n42, 178n44; and GMOs, 147–48, 188; and Golden Rice, 150; iron deficiency anemia in children in, 114; national fortification projects in, 46; tribal health and diets in, 22; wheat flour mandatory fortification standards in, 93. *See also* *specific countries*

AFTA, 74

AGRA (Alliance for a Green Revolution in Africa), 14, 15–16, 177n39, 188n9

agricultural labor force, feminization of, 6–7, 15, 174nn11–13, 178n43

agricultural research and development, 140–49, 154, 156–57, 188nn8–11. *See also* biofortification; Golden Rice; Green Revolution

agricultural surpluses. *See* wheat exports

agrofood studies, 9, 10, 163, 168, 175n19

Alliance for a Green Revolution in Africa (AGRA), 14, 15–16, 177n39, 188n9

alternative agrofood movements, 169–71

Altieri, A., 15

anemia. *See* iron deficiency anemia (IDA)

anti-GMO movement, 150–51, 153, 160

Apple, Rima, 35, 136, 137, 163, 186n2

APTINDO (Indonesian Association of Wheat Flour Producers), 85, 87, 91, 93–95, 184–85n5

Archer-Daniels-Midland, 182n18

Asian Development Bank (ADB): biofortification projects of, 146, 157; fortification projects of, 40, 44, 47, 52, 68–69; and gender mainstreaming imperative, 58; on malnutrition as economic loss, 52

asymmetrical mother-child dyad, 20, 27, 31–34, 136

Avakian, Arlene Voski, 6

Ayahbunda, 120–22, 186n4

baby food: commercial market for, 117–19; complementary food (CF) and supplementary food (SF), 114–16, 124, 127, 130, 133, 187n7; expert discourse on, 115, 116, 130–31; homemade baby food in Indonesia, 116, 124; international opposition to commercial baby food, 131–32; making of "smart" baby food, 114–19; marketing and advertising of, 119–23, 126, 129–35, 137, 167, 186n4; nutritional claims for commercial baby food, 120–23, 125–31, 135, 167; and nutritionalized self, 134–38; and nutritional needs of infants, 111, 114–15; and nutritionism, 111–14; and scientization of motherhood, 112–14

BAFF (Business Alliance for Food Fortification), 47, 53–54, 173n7

Belasco, Warren, 31, 159, 189n19

benevolent biotechnology, 146–49, 158, 159–60

Berdikari Flour Mill, 84, 85, 87, 88, 89, 91

Bill and Melinda Gates Foundation, 14, 41, 145, 146, 177n37, 177n39, 188n9, 188n11

biofortification: analysis of, in agrofood studies, 10; beginning of concept of, 139, 188n3; as benevolent biotechnology, 146–49, 158, 159–60; challenges of, 143; and combination of agriculture and nutrition, 140–46, 149, 157; definition of, 2, 139; funding for, 145–46, 177n39; and GMOs, 3, 146–51, 153, 155, 159, 160, 173n4; of Golden Rice, 2, 17, 114, 139, 144–46, 149, 153–58; and international development, 145–46, 149; moral politics of, 158–61; promotion of, 139–40, 143–46, 149; public-private partnership for, 5–6, 173n7; social appeal of, 149. *See also* fortification; Golden Rice

biological citizenship, 135–36

biological victimhood of women, 8, 59–61, 84, 134–38, 163, 165, 170, 183nn22–23, 189n19

biopower, 11, 63, 137–38, 176n31

Bogasari Flour Mill, 81, 82, 85–91, 104–5, 183n7, 184n4